Plenum Series on Demographic ods and Population Analysis

:ditor: Kenneth C. Land, *Duke University, Durham, North Carolina*

ICED TECHNIQUES OF POPULATION ANALYSIS
. Halli and K. Vaninadha Rao

TICAL THEORY OF BIOLOGICAL POPULATIONS
. Lotka

UITIES IN SOCIOLOGICAL HUMAN ECOLOGY
y Michael Micklin and Dudley L. Poston, Jr.

IG POPULATION GROWTH: An Insider's Perspective on the on Movement
arkavy

MOGRAPHY OF HEALTH AND HEALTH CARE
Pol and Richard K. Thomas

L DEMOGRAPHY
Smith

IOLD COMPOSITION IN LATIN AMERICA
De Vos

IOLD DEMOGRAPHY AND HOUSEHOLD MODELING
Evert van Imhoff, Anton Kuijsten, Pieter Hooimeijer, van Wissen

NG MULTIGROUP POPULATIONS
hoen

ULATION OF MODERN CHINA
Dudley L. Poston, Jr. and David Yaukey

R OF POPULATION DYNAMICS
Namboodiri

on Order Plan is available for this series. A continuation order will bring ach new volume immediately upon publication. Volumes are billed only upon ent. For further information please contact the publisher.

Analytical Theory of Biological Populati

The
Met

Series

ADVA
Shiva

ANAL
Alfred

CONT
Edited

CURB
Populat
Oscar H

THE DI
Louis G

FORMA
David P

HOUSE
Susan M

HOUSE
Edited b
and Leo

MODEL
Robert S

THE PO
Edited b

A PRIM
Krishnan

A Continua
delivery of
actual ship

Analytical Theory of Biological Populations

Alfred J. Lotka

Translated and with an Introduction by

David P. Smith
University of Texas
Houston, Texas

and

Hélène Rossert
AIDES Fédération Nationale
Paris, France

Plenum Press • New York and London

Library of Congress Cataloging-in-Publication Data

Lotka, Alfred J. (Alfred James), 1880-1949.
[Théorie analytique des associations biologiques. English]
Analytical theory of biological populations / Alfred J. Lotka ; translated and with an introduction by David P. Smith and Hélène Rossert.
p. cm. -- (The Plenum series on demographic methods and population analysis)
Includes bibliographical references and index.
ISBN 0-306-45927-2
1. Demography--Mathematical models. I. Title. II. Series.
HB849.51.L67313 1998
304.6'01'5118--dc21 98-41063
CIP

ISBN 0-306-45927-2

A Division of Plenum Publishing Corporation
233 Spring Street, New York, N.Y. 10013

http://www.plenum.com

10 9 8 7 6 5 4 3 2 1

Printed in the United States of America

For Ken

Preface

In the 50 years that have passed since Alfred Lotka's death in 1949 his position as the father of mathematical demography has been secure. With his first demographic papers in 1907 and 1911 (the latter co-authored with F. R. Sharpe) he laid the foundations for stable population theory, and over the next decades both largely completed it and found convenient mathematical approximations that gave it practical applications. Since his time, the field has moved in several directions he did not foresee, but in the main it is still his.

Despite Lotka's stature, however, the reader still needs to hunt through the old journals to locate his principal works. As yet no extensive collections of his papers are in print, and for his part he never assembled his contributions into a single volume in English. He did so in French, in the two part *Théorie Analytique des Associations Biologiques* (1934, 1939). Drawing on his *Elements of Physical Biology* (1925) and most of his mathematical papers, Lotka offered French readers insights into his biological thought and a concise and mathematically accessible summary of what he called *recent contributions* in demographic analysis. We would be accurate in also calling it *Lotka's contributions* in demographic analysis.

Lotka had intended to write a third part to the *Théorie Analytique* drawing heavily on the *Elements*, but abandoned the project after completing Part 2. In the *Analytical Theory of Biological Associations* we present an English translation of the two parts he completed.

Readers with a background in mathematical demography will find much of the *Analytical Theory* reasonably familiar, since mathematical demography cannot be taught without drawing heavily on Lotka, both for essential formulas and for the many practical applications that breathe life into them. A few of the problems Lotka introduced, principally involving logistic population models, are no longer current, but for the rest — the renewal equation and stable population theory, observed and intrinsic rates, relations between various demographic measures, and specific problems associated with reproductivity and deaths — the *Analytical Theory* stands as a valuable source document for modern work.

In completing this volume we are particularly indebted to the staffs of the libraries of the University of Texas School of Public Health and Rice University for their help in locating Lotka's many published papers, and to Beatrice Keyfitz for the insight, immediately evident, that we should have those in hand before undertaking the translation of a book that itself drew heavily from them.

David P. Smith
Hélène Rossert

Houston, Texas

Contents

INTRODUCTION

Lotka and the Théorie Analytique

Alfred James Lotka[*] was born in Lemburg, Austria (now Lvov, Ukraine) March 2, 1880 of American parents, Jacques and Marie (Doebely) Lotka. He grew up in France, then earned his B. Sc. Degree at Birmingham University in 1901. Over the next year he did graduate work in chemistry at the University of Leipzig, where he began to consider relations between chemical and biological systems in the context of thermodynamics, an insight he drew from the lectures of Wilhelm Ostwald. He came to the United States in 1902, for the next six years working as an assistant chemist for the General Chemical Company. While there (1907) he published two papers, "Relation between birth rates and death rates," and "Studies on the mode of growth of material aggregates," in which he introduced relations between birth and death rates, the age distribution and population increase under stability, with the remark that closed populations having approximately constant fertility and mortality would tend toward that state. That insight became the basis of much of his life's work in biology and demography.[†] From

---

[*] The biographical notes included here are from the obituary notices by Louis Dublin (*Journal of the American Statistical Association*, 1950, *v. 45*, pp. 138–139), Frank Lorimer (John A. Garraty and Edward T. James, editors, *Dictionary of American Biography, Supplement 4: 1946–1950*, pp. 505–506. American Council of Learned Societies, 1974), and Frank W. Notestein (*Population Index*, 1950, *v. 16*, pp. 22–29). The latter includes a bibliography of Lotka's papers, as does the Dover edition of Lotka's *Elements of Mathematical Biology* (Dover, New York, 1956). Much additional information about Lotka's life and work, drawn in part from the Alfred James Lotka Papers at Princeton, is given in Sharon E. Kingsland, *Modeling Nature* (University of Chicago Press, Chicago, 1985), from which we have taken our comments on Lotka and Raymond Pearl.

[†] Lotka, Alfred J. "Relation between birth rates and death rates," *Science*, 1907, *v. 26*, pp. 21–22. Reprinted in David P. Smith and Nathan Keyfitz, *Mathematical Demography:*

1908 to 1909 he studied at Cornell, where he received an M. A. in physics. He then spent a year as a patent examiner in the U. S. patent office, and two years as an assistant physicist in the U. S. Bureau of Standards. While at the Bureau he published "A problem in age distribution," with F. R. Sharpe,* which moved beyond relations under stability by introducing the renewal function in the form familiar to demographers

$$B(t) = \int_0^\infty B(t-a)\, p(a)\, m(a)\, da \qquad (134)^\dagger$$

In the expression, $B(t)$ represents births for one sex at time t, $p(a)$ is the survival probability from birth to age a, and $m(a)$ is the fertility probability at age a with respect to births of the same sex as the parent. Lotka identified the unique real root ρ of the expression as the solution to

$$1 = \int_0^\infty e^{-ra}\, p(a)\, m(a)\, da \qquad (137)$$

and recognized it as the rate of increase to which the distribution would ultimately converge.

During the next three years Lotka served as editor of the Scientific American Supplement, and completed his D. Sc. at Birmingham University. He afterwards returned to the General Chemical Company, remaining there until 1919. In 1922 Raymond Pearl secured funding to bring him to Johns Hopkins University, where he spent two years writ-

---

Selected Papers, Springer–Verlag, Berlin, 1977, pp. 93–95. Lotka, Alfred J. "Studies on the mode of growth of material aggregates," *American Journal of Science*, 1907, *v. 24*, pp. 199–216, 375–376.

* Sharpe, F. R. and Alfred J. Lotka. "A problem in age distribution," *Philosophical Magazine*, 1911, *v. 21*, pp. 435–438. Reprinted in David P. Smith and Nathan Keyfitz, *Mathematical Demography: Selected Papers*, Springer–Verlag, Berlin, 1977, pp. 97–100.

† Except where noted the expression numbers follow Lotka's numbering in Part 2 of the *Analytical Theory of Biological Associations*.

ing the *Elements of Physical Biology*,[*] published in 1925. The year 1922 also saw publication of "The stability of the normal age distribution,"[†] in which he offered a graphic proof of the convergence of populations to stability under fixed fertility and mortality, which stood until Feller's rigorous mathematical proof was published in 1941.[‡]

For the remainder of his working life Lotka was with the Metropolitan Life Insurance Company, retiring in 1947. After retirement he had hoped to put his contributions to demography in book form in English, but he died in 1949 without completing it.

Fortunately, Lotka left us with a version of his planned book in French, the *Théorie Analytique des Associations Biologiques*, published in *Actualités Scientifiques et Industrielles* (Hermann, Paris) in two parts in 1934 and 1939. The *Analytical Theory of Biological Associations* is the English translation of that work. Part 1, *Principes* [*Principles*], completed in 1934, is an introduction to the mathematical treatment of biological systems, and draws much of its focus from the early chapters of the *Elements*. Part 2, the *Analyse Démographique avec Application Particulière à l'Espèce Humaine* [*Demographic Analysis with Specific Application to the Human Species*], published in 1939, presents his major contributions to mathematical demography. There was to have been a third volume incorporating much of the mathematical analysis of the *Elements*, but that was never completed.

As most demographers are not familiar with Lotka's contributions in evolutionary biology, a comment on his focus in Part 1 may be helpful. We will take things somewhat out of Lotka's order.

To begin on familiar ground, the biological processes that concern demographers have to do with a single species, most commonly our-

---

[*] Lotka, Alfred J. *Elements of Physical Biology*, Williams and Wilkins, Baltimore, 1925. Reprinted with corrections and a bibliography of Lotka's publications as *Elements of Mathematical Biology*, Dover, New York, 1956.

[†] Lotka, A. J. "The stability of the normal age distribution," *Proceedings of the National Academy of Sciences*, 1922, *v. 8*, pp. 339–345.

[‡] Feller, W. "On the integral equation of renewal theory," *Annals of Mathematical Statistics*, 1941, *v. 12*, pp. 243–267.

selves, and with the implications fertility and mortality rates have for population and family composition, and for the social relations arising from them. That analysis can be embedded in a more general treatment that incorporates inter-species relations, and, as Lotka emphasized, needs to be so embedded for every species except ourselves. We move to a third level when the time periods that concern us are long enough that alterations in the inanimate environment, and the effects of mutations on the composition of each species, also need to be taken into account.

All of these aspects can be incorporated in a model of the form

$$\left.\begin{aligned}
\frac{dX_1}{dt} &= F_1(X_1,X_2, \ldots ,X_n;P,Q) \\
\frac{dX_2}{dt} &= F_2(X_1,X_2, \ldots ,X_n;P,Q) \\
&\cdots \\
\frac{dX_i}{dt} &= F_i(X_1,X_2, \ldots ,X_n;P,Q) \\
&\cdots \\
\frac{dX_n}{dt} &= F_n(X_1,X_2, \ldots ,X_n;P,Q)
\end{aligned}\right\} \qquad (4)^*$$

where the terms in X_i denote the various species in the system, P are salient aspects of the inanimate environment, such as topography and climate, as they affect each species, and Q are relevant biological characteristics of the different species and their individual members, which we take as fixed in the short term but susceptible over time to evolutionary change. The terms in the model thus relate the changes in number and character of each given species dX_i/dt to changes in its niche, including changes both in the physical environment (P) and in itself and the other species that interact with it (Q). Without abusing the terms of the model, we can also shift our perspective from individuals and spe-

---

* Part 1, p. 31.

cies to the *energy content* each species represents, which is to say, the thermodynamics of the system.

For the simplicity of its notation, the model is dramatic in its sweep. Lotka did not expect its terms to be quantified with any precision given the scant information then available, but all aspects of biological evolution, including the evolution of the human mind, fall within its scope. The pieces Lotka contributed range from his appreciation of the chemical and energy balance of the earth to his insights into the mathematics of species interaction, the biological role of the senses and human consciousness, the human economy, and the demography of human populations. It is still possible to be challenged by the breadth of Lotka's investigations.

Part 1 begins rather more simply, with a comparison between the elementary chemical relation

$$2H_2 + O_2 \Leftrightarrow 2H_2O \qquad (1)^*$$

and the biological relation we have already noted

$$\frac{dX_i}{dt} = F_i(X_1, X_2, \ldots, X_n, P, Q) \qquad (2)^\dagger$$

The two expressions bring out the fundamental link but also the evident differences between models for chemical and biological systems. In both cases the analyses are of *systems*, as the individual components provide only part of the information we require. They differ in that biological systems are observed at a different level than are chemical systems and involve entities that are only relatively homogeneous. Their evolution is also essentially open ended, has direction, and for our species involves mind and volition, all of which are troublesome characteristics for the theoretician who demands scientific precision.

---

* Part 1, p. 4.

† Part 1, p. 7.

Lotka was that theoretician, and his aim in much of Part 1 was to introduce readers to the difficulties these distinctions present. After establishing the distinction between studying systems and their individual components, Chapter 1 considers the nature of energy transformations among biological species and the incomplete homogeneity within species that arises through reproduction, from which the essentially open endedness of evolution derives. Lotka also briefly notes the distinction between levels of observation in chemical systems, where individual units (atoms and molecules) are usually invisible to us, and biological systems, where the units (biological organisms) are what we most readily observe.

Chapter 2 addresses the nature of time, which must be irreversible for us to make sense of evolution, but need not be in classical mechanics, and cannot be shown to be irreversible, or other than irreversible for that matter, by any tests we able to construct. Absent that proof, we find ourselves proceeding on the evidence such as it is, with perhaps more faith than we recognize.

Chapter 3 offers a further comment on irreversibility as it applies to biological organisms, considered from the perspective of living organisms as transformers of matter, and on problems imposed by the scale at which the transformations are observed. Here also we do not know if an element of randomness or capriciousness, as would be imposed by changes in the direction of time, enters into our models once we are at scales where the uncertainty principle precludes exact measurement, or if something of a random or capricious nature might arise by an imposition of human will to which laws of thermodynamics do not apply. Of necessity, we proceed on the understanding that we do not introduce effects we cannot define.*

Having introduced these issues, which touch on the foundations of the mathematical treatment of biological processes, Lotka was ready to present the model in its general form in Chapter 4, and to demonstrate the insights that can be derived from it even when it takes a very simple

---

* To be fair, Lotka himself wavered as regards the concept of free will (pp. 24, 28), which he was reluctant to disregard, but which, to quote him in a not dissimilar context (p. 26), "introduces a new term into our discussions without introducing new facts."

form, holding the parameters P and Q constant. The model then expresses the types of dependence that may exist among species at one point in time, from elemental predator–prey models to the functioning of human economies. Lotka did not offer it as an illustration, but would certainly have included his most famous example in his intended third volume. For a simple predator–prey model, comprising two species in which frequency of contact is a function of the density of the species, the general model assumes the form of the familiar Lotka–Volterra equation:*

$$\left.\begin{aligned}\frac{dX_1}{dt} &= X_1(\varepsilon_1 - \gamma_1 X_2)\\ \frac{dX_2}{dt} &= X_2(\gamma_2 X_1 - \varepsilon_2)\end{aligned}\right\}$$

where for the prey $(X_1), \varepsilon_1$ is the birth rate (or, the birth rate less the rate of death for causes not associated with predation), and γ_1 is the death rate due to predation. For the predator (X_2), γ_2 is the birth rate associated with successful predation (or, the difference between the birth and death rates associated with predation), and ε_2 is the death rate for other causes. The most interesting applications of the model, because they are the only sustainable ones, are those that lead to stationary equilibria, but these are by no means the only outcomes the system can have. Depending on the magnitudes of the separate coefficients, the system may or may not converge from an arbitrary initial configuration to a stable or cyclically stable state. It can also end with the prey and predator vanishing, or with only the prey flourishing or perhaps increas-

---

* Lotka, *Elements.*, pp. 92–93; Volterra, Vito, *Leçons sur la Théorie Mathématique de la Lutte pour la Vie*, Gauthier–Villars, Paris, 1931, p. 14. See also Volterra, Vito, "Variazioni e fluttuazioni del numero d'individui in specie animali conviventi," *R. Comitato Talassografico Italiano, Memoria*, 1927, *v. 131*, pp. 1–142. [Available in English in Vito Volterra, "Variations and fluctuations in the numbers of coexisting animal species," translated by Francesco M. Scudo and James R. Zeigler. Pp. 65–236 in Scudo and Zeigler, editors, *The Golden Age of Theoretical Ecology*: *1923–1940,* Springer–Verlag, Berlin, 1978.]

ing indefinitely in the predator's absence. It will be recognized that the various outcomes are not descriptive of known species to a high degree of accuracy, but that is more than Lotka required and he pointedly said so to critics who saw it as an endpoint rather than a starting point for the analysis of interacting species.

The Lotka–Volterra equation carries us beyond Part 1 of the *Analytical Theory*, and is included here to help the reader appreciate the insight that can be derived from even elementary models. After presenting the general model, Lotka addressed some of the implications it has for the case of multiple species interacting with each other, for which the reader may see pp. 57–63, and 77–99 of the *Elements*. He then turned briefly to the special case of a single species considered in isolation, which allowed him to introduce the exponential and logistic population models considered in detail in Part 2 of the *Analytical Theory*: *Demographic Analysis with Specific Application to the Human Species*.

Lotka begins the *Demographic Analysis* with a note on the distinctiveness of our own species, the complexity of the demographic relations that affect us, and the abundance of data we have for ourselves — conditions which compel us to bring order to our understanding of ourselves where we can through the concise language mathematical modeling offers. In doing that, the two problems that will most affect us are the sheer number of variables that may enter and the probabilistic nature of the relations between them. Lotka addresses these both in the general organization of the book and the extensive use he makes of distribution means and variances to reduce the complexity of the relations he develops. The reader might note in passing the large proportion of footnotes that refer to Lotka's own papers. During his lifetime the field was very largely his.

Proceeding in Lotka's order, Chapter 2 briefly defines terms, including those for the birth, death and increase rates, and in Lotka's expression (9) establishes the fundamental relation between the total population $N(t)$ at time t, births $B(t-a)$ earlier in time, and the probability of survival to age a, $p(a)$:

$$N(t) = \int_0^{\omega} B(t-a)\, p(a)\, da \tag{9}$$

using ω to represent the oldest age to which anyone survives. From this follow the series of relations by which the number and proportion in a given age interval are found, and from which vital rates are computed. In the special case of a population with a constant proportion $c(a)$ by age, that is, a *stable* population, the various rates are necessarily also constant, and numbers of births and deaths are either constant or change at a constant rate over time [Lotka's expressions (21)–(44)].

Lotka then considers the shapes of different stable age distributions, demonstrating that the mean age is fairly robust to very substantial changes in the rate of natural increase and the proportions at the youngest and oldest ages. Lotka's treatment includes development of the relation between the rate of increase r and the mean age of the stable population A_r [expression (48)], and by way of A_r, the relation between r and the stable birth rate b [expression (64)].

At various points, beginning from expression (45), Lotka also introduces what has become one of the most powerful tools in demographic analysis, the series expansions for various integrals from which a number of quite good approximate formulas are derived. The key in each instance has been recognizing when most of the information about a distribution is contained in its first few moments or cumulants, permitting higher order terms to be dropped. The critical series expansions with which Lotka worked are provided as expressions (47)–(49) and (100)–(108), and as Appendix 1. As Lotka established, both distribution moments and cumulants enter naturally in demographic analysis, the moments being associated with characteristics of stationary populations and cumulants being associated with characteristics of stable populations having rates of increase $r \neq 0$.

Besides the approximate formulas he developed from series expansions, the tedium of estimating moments and cumulants before computers led Lotka to examine the theoretical basis for certain empirical expressions that required little math. One example is his derivation (pp. 75–81) of a useful approximation for the first cumulant of the survival distribution, the mean age of the life table population [from (49), (75)],

$$\lambda_1 = \frac{\int_0^{\omega} a\,p(a)da}{\int_0^{\omega} p(a)da} = \frac{L_1}{L_0} = A_0$$

as the value λ_1 satisfying $\lambda_1 = \mathring{e}_{\lambda_1}$. This relation is shown to be exact under De Moivre's hypothesis, whereby the survival probability $p(a)$ declines linearly with age, and to hold to a close approximation for contemporary United States. and European populations. It is still rather good. For higher order moments and cumulants, Lotka suggested substituting terms from similar populations for which they had already been worked out, and to buttress that suggestion, he presented tables of moments and cumulants for selected places to confirm their stability.

After introducing the simple approximation for the mean age A_0, Lotka considers another empirical problem (pp. 86–88), the determination of the constant $0 \le \alpha \le 1$ satisfying

$$\frac{\alpha}{b} + \frac{1-\alpha}{d} = L_0$$

Lotka finds for stable populations

$$\alpha \doteq 1 - \frac{L_1}{L_0^2}$$

The expression yields numerical estimates at least tolerably in agreement with a contemporary (circa 1870) approximation attributed to William Farr, $(1/3)\frac{1}{b} + (2/3)\frac{1}{d} = L_0$. Here and in earlier expressions the reader will recognize L_0 as the life expectancy at birth. The discovery of formal relations underpinning useful empirical formulas is one of the small pleasures of mathematical investigation. Lotka did it well.

The remainder of Chapter 2 addresses relations between births, deaths, and population size when one of the quantities changes over time. Lotka found the general result, which follows from the concentration of births and deaths in particular age ranges, that whatever law is established for one of the three quantities holds to a close approximation for the other two as well with appropriate adjustment of the time origin [expressions (100)–(114)]. The equivalence is exact for stable populations, already considered [expressions (21)–(35)]. In expressions (117)–(133) the formulas are applied to a population whose size follows a logistic distribution, for which the numbers of births and deaths, and vital rates, are found to also approximate the logistic. The result is intuitive, and with the increasing concentration of mortality in old age holds somewhat more generally now than it did in Lotka's time, but the forms of the distributions in no way convey that insight when a common time origin is imposed, as lags between births, population size, and deaths must be accommodated through the high order terms of series expansions. Incorporating lags by recentering each distribution at its approximate mean allowed Lotka to recover the intuitive solution.

The principal use of the logistic was as a forecasting tool, for which it was already losing favor when Lotka wrote the *Analytical Theory*. The more important result he derived was in the application of the formulas to historical data. For the United States, the model reasonably tracked fertility and mortality into the 1930s and identified the secular upturn in mortality to be expected after 1940 if fertility remained low, both substantial results for so elementary an exercise. They were, however, its limit. Quite apart from the impact of the Second World War and the subsequent baby boom, the value of the logistic for population projection has become marginal in an age richer in demographic methods and much richer in data and computing technology than was Lotka's.

Having presented general relations between populations and vital events in Chapter 2, and having introduced the characteristics of stable populations, Lotka was ready in Chapters 3 and 4 to develop stable population theory. Chapter 3 is given to an examination of the roots of the fundamental equation

$$1 = \int_0^\infty e^{-ra}\, p(a)\, m(a)\, da \tag{137}$$

derived from expression (134) by substitution of the relation under stability

$$B(t) = B_0 e^{rt} \tag{29}$$

For the solution to (137) Lotka makes use of the series expansion [expressions (156) and (160)] for the interval between generations T_r

$$\int_0^{\infty} e^{-ra} p(a) m(a)\, da = R_0 e^{-rT_r} = R_0 e^{-r(\mu_1 - \mu_2 \frac{r}{2!} + \cdots)} = 1$$

from which

$$\mu_1 r - \mu_2 \frac{r^2}{2!} + \cdots - \ln R_0 = 0 \tag{162}$$

an expression involving only r, the mean fertile age of the stationary population μ_1, its variance μ_2, and the net reproduction rate R_0. The concentration of human fertility about its mean and the near symmetry of the fertility distribution allow higher cumulants to be ignored. For human populations the error is small even when the term in μ_2 is dropped. The estimation of the period of the first complex root and of the intrinsic birth rate b is only slightly more involved.

Chapter 4 completes the development of stable theory by demonstrating the increasing overlap of births across different generations of parents, through which the initial distinctiveness of a population's age structure is lost, and by determining the constant terms Q_s representing the stable equivalents to the initial population associated with each of the intrinsic roots r_s. To develop the solution, Lotka introduces the new term P_s, defined as

$$P_s = \int_0^{a_2} e^{-r_s t} B_1(t)\, dt \tag{214}$$

On substituting terms in $B(t)$ and $B(t-a)$ for $B_1(t)$ and introducing the series expansion

$$B(t) = \sum Q_j e^{r_j t} = Q_s e^{r_s t} + \sum_{-s} Q_j e^{r_j t} \tag{135}$$

and the similar expansion for $B(t-a)$, Lotka is able to express P_s in terms of Q_j and r_j. He is then able to show that terms of the series involving Q_{-s} and r_{-s} cancel, as we expect at least for the principal term Q_ρ since populations sharing the same stable configuration may or may not have roots other than the intrinsic rate of increase ρ in common. The relation becomes

$$Q_s = P_s \Big/ \int_0^{a_2} a e^{-r_s a} \varphi(a)\, da \tag{230}$$

or more simply, since we recognize the denominator of the expression as the mean age at childbearing in the stable population A_r (or A_{r_s})

$$Q_s = P_s / \mathrm{A}_{r_s}$$

The expression P_s had been introduced by R. A. Fisher* as the Total reproductive value of a population, essentially the backward projection of fertility accruing to an observed population to find the size of the birth cohort that would be reproductively equivalent to it. For stable theory it is not the equivalent population to the end of the reproductive age interval P_s that we require but its annual births, analogous to $B(t)$, found

---

* Fisher, R. A. "The actuarial treatment of official birth records," *Eugenics Review*, 1927, *v. 19*, pp. 103-108; and *The Genetical Theory of Natural Selection*, pp. 25–30, Dover, New York, 1958 (1930). See also Fisher's correspondence with Lotka, *Eugenics Review*, 1927, *v. 19*, pp. 257-258.

as P_s / A_{r_s}. Although aware of Fisher's work, Lotka did not see its link to his own analysis or recognize that P_s could be given an intuitive interpretation.

Having completed his presentation of stable theory, Lotka uses Chapter 5 to contrast measures of population change, with emphasis on approximations to the net reproduction rate R_0 based on births between two generations, and on observed and life table ratios of persons in different age intervals. The latter is familiar to us through Thompson's Index, relating child–woman ratios in real populations to those in a suitable life table:

$$J = \frac{\int_0^5 c(a,t)\,da}{\int_{15}^{45} c_f(a,t)\,da} \div \frac{\int_0^5 p_f(a)\,da + s\int_0^5 p_m(a)da}{\int_{15}^{45} p_f(a)\,da}$$

where s is the sex ratio at birth and f, m identify gender. The denominators span age intervals appropriate to the population (we have used the contemporary United States). Lotka's closing comments on the distinction between R_0 and the intrinsic rate of increase ρ will be familiar to modern readers. Lotka's remark (p. 157) that the intrinsic rate of increase ρ would be below the crude rate r in a population whose family sizes were declining continues to hold for the United States.

In Chapters 6 through 8 Lotka moves from characteristics of populations to characteristics of the component individuals and families, involving family sizes in Chapters 6 and 7, the influence of parental mortality in Chapter 7, and the disappearance of family surnames as a function of family size in Chapter 8. Orphanhood receives particular attention: in 1921 11 percent of English and Welsh children under age 16 had lost one parent, in part because of the World War. For the United States in 1930 the figure may have been near 8 percent.

The more interesting problem Lotka examines, attractive in part because it was initially formulated correctly and answered incorrectly, is the relation between the distribution of families by size and the probability that particular family traits (here, surnames) will disappear. In his treatment, Lotka extends the distribution function for the number of offspring

$$f(x) = \pi_0 + \pi_1 x + \pi_2 x^2 + \cdots \qquad (314)$$

across generations by the iterations $f_s(x) = f\{f_{s-1}(x)\}$ by which he is able to diagram (Figure 20, p. 185) the approach of the cumulative extinction probability to an asymptotic value $0 \leq \xi \leq 1$. The probability is found to be determined by the proportion of individuals having no offspring π_0, the net reproduction rate R_0, and in a less obvious way by the ratio of small to large families, since the former are necessarily at greater risk of passing out of existence.

Lotka concludes the *Analytical Theory* with a note on the need for a probabilistic representation of demographic phenomena, which in his day required urns with varying distributions of black and white balls, or disks that could be spun, marked to conform to probabilities for different events. As he would have appreciated modern computers, the reader can appreciate the remarkable accomplishments of the generations of scientists such as himself who lacked them.

A word should be said about origins. Part of stable population theory was originally developed by Leonard Euler, in two publications that were better known on the continent than in the United States.* A general

---

* Euler, Leonard. "Recherches générales sur la mortalité et la multiplication du genre humaine." *Histoire de l'Académie Royale des Sciences et Belles Lettres*, 1760, *v. 16*, pp. 144–164.; and Süssmilch, Johann P. *Die göttliche Ordnung*. 1798 (1761), *v. 1*, pp. 291–299. The articles will be found in English translations in Euler, "A general investigation into the mortality and multiplication of the human species," translated by Nathan and Beatrice Keyfitz. *Theoretical Population Biology*, 1970, *v. 1*, pp. 307–314; and Smith, David P. "An Euler contribution to Süssmilch's göttliche Ordnung," with a translation by Nathan Keyfitz, *Theoretical Population Biology*, 1977, *v. 12*, pp. 246–251. The first paper and most of the second are reprinted in Smith, David P. and Nathan Keyfitz. *Mathematical Demography: Selected Papers*. Springer–Verlag, Berlin, 1977, pp. 79–91. The different pieces of Euler's contribution were first assembled by E. J. Gumbel ["Eine Darstellung statistischer Reihen durch Euler," *Jahresbericht der Deutschen Mathematiker Vereinigung*, 1917, *v. 25*, pp. 251–264], cited by Lotka.

comment on Euler by George Simmons[*] is appropriate: "Euler's researches were so extensive that many mathematicians try to avoid confusion by naming equations, formulas, theorems, etc., for the person who first studied them after Euler." Lotka was one. For his part he was annoyed that his own contributions were unappreciated by Robert Kuczynski, and also by Volterra, who credited him with independently developing the Lotka–Volterra equation, but not with his more general insights.[†]

Lotka was himself less than effusive in his comments on Euler's remarkable papers. He dismissed Euler's numerical example (in Süssmilch) of the eventual stabilization of a population beginning from two parents at age 20 giving birth to sets of twins at ages 22, 24, and 26, as "the first crude approach by Euler (based on highly unrealistic assumptions, and quite inapplicable to actual statistical data) ..."[‡] Euler, who plainly knew the implications of his example, might fairly have responded in the same tone Lotka reserved for Kuczynski and Volterra. Again with respect to attribution, Lotka's mathematical expressions for

---

[*] Simmons, George F. *Differential Equations with Applications and Historical Notes*, McGraw–Hill, New York, 1972, p. 86.

[†] "Le Dr. Lotka, dans son Ouvrage *Elements of Physical Biology* (New York, 1925) qui comprend de nombreuses applications des mathématiques à des questions de chimie et de biologie a envisagé le cas de deux espèces et donné une représentation géométrique des variations ainsi que la période des petites fluctuations. [Dr. Lotka, in his work *Elements of Physical Biology*, which includes numerous applications of mathematics to questions in chemistry and biology, envisaged the case of two species and gave a geometric representation of the variations and of the periodicity of small fluctuations.]" Volterra, Vito. *Leçons sur la Théorie Mathématique de la Lutte pour la Vie*. Gauthier–Villars, Paris, 1931, p. 4.

[‡] Lotka, Alfred J. "A historical error corrected." *Human Biology*, 1937, *v. 9*, pp. 104–107. The quotation is from page 107. A similar comment, citing Euler's 1760 article, is given in the note on p. 886 of Lotka, 1928, *loc. cit.* The citation should have been to Euler's contribution in Süssmilch, which suggests that Lotka was working strictly from Gumbel, *loc. cit.*, and had not yet located the original sources.

relations between births, deaths and natural increase in stable populations are largely given or anticipated in Euler's 1760 paper. It would have been appropriate for Lotka to have recognized Euler in this context in the *Analyse Démographique*, but he let the opportunity pass. Lotka's reputation survives this slight both by the substance of his own contributions and by the chasm between Euler's appreciation of stable theory and the pieces he put into print. It is through Süssmilch, who left nothing he thought substantive unpublished, that we learn Euler investigated the convergence of populations to stability at all.

Finally, we remark for readers that apart from familiar differences between French and English grammar, our translation from the original is essentially literal, as Lotka used much the same terminology and more or less the same writing style in both languages. We have made some adjustments for particular word preferences Lotka had in one language but not the other, and at a handful of points we have lifted a critical phrase from one of Lotka's English publications or otherwise strayed from his specific wording where a direct translation from the French was close to unintelligible. Occasionally Lotka seems to have bulldozed his way from English through to French, especially when the source text had itself not been written clearly. For the most part, however, Lotka's French was competent if pedestrian (a comment that is probably an oxymoron in French but is not in English), and does not present exceptional difficulties for the translator.

The writing was, however, hurried. Entries in the Table of Contents did not quite match chapter subheadings, which is corrected here; on two or three occasions *rate* is used to mean *probability*, not corrected; and throughout there is a repetitiveness favoring particular words and phrases that Lotka would probably have avoided if he had given the text more time. We have allowed that to stand, remaining faithful as best we could to the French Lotka used, even though it was not as well as he could have done, and reads in the original and in translation as a slightly unfinished effort.

PART I

Principles

CHAPTER 1

The order of ideas which has given birth to the modern theory of organic evolution is by its nature essentially quantitative. We know that Darwin, and Wallace as well, were drawn to the enunciation of the principle of survival of the fittest through reflection on the problem first posed by Malthus: how is the number of living organisms maintained within the limits we in fact observe? By a rather singular coincidence this aspect of the problem of evolution, which, it would seem, would have been the first to attract the attention of biologists predisposed to mathematical analysis, has only quite recently received their serious attention. A certain school, it is true, profiting by the work in genetics of Gregor Mendel among others, has been occupied for some years with biometrical anal-ysis applied to questions of survival and reproduction in their relation to the problem of organic evolution. But, in their research, the disciples of this school have limited themselves almost entirely to the discussion of the characteristics of a single species and the consequences of these characteristics as they affect its survival. The interaction of diverse species among themselves and with their habitat has received at most passing and incidental consideration by these authors. Ecologists, by contrast, have been content almost entirely with empirical studies on this subject.

There thus remains a great lacuna in the biometrical sciences. Today we see a group that is still relatively small occupied in filling it. It would seem that we are witnesses at the birth of an entirely new branch of biological science, which one could call *general demology*. It is concerned with the analytical study of aggregations formed by populations of diverse biological organisms.

There are very special considerations that lend this branch of biometry a pre-eminent importance. One cannot insist too strongly on the fact that the evolution of a system comprising a certain number of biological species ought to be conceived as a whole. It is not this species or that one which evolves, but the system itself as an entire system. Alongside certain species which have acquired highly perfected organs of attack, there exist, and indeed must exist, other species whose defense consists in large part in elevated and often even prodigious fertility. I say that such other species must exist, for without them the former would not survive indefinitely for lack of food. It is thus mutually dependent groups which evolve, and which evolve as groups, and every study of the fundamental problem of evolution that ignored this fact would be condemned in advance to failure. It would be vain, for example, to pose the question: does the evolution of a species tend toward more and more perfect development of its means of defense against attacks by other species, or does it tend rather toward a level of fertility powerful enough to guarantee the survival of a sufficient number even despite such attacks? The fact is that both of these tendencies are realized. The question is badly posed. The direction of the process we call *evolution* could not be meaningfully defined in terms of a single species.

In these reflections it is very useful to profit from the more advanced understanding we possess of the evolution of certain inanimate systems. The direction of this evolution can be defined in a precise fashion, but, to be sure, always in terms of certain characteristics of the *entire* system. Given for example a sealed vessel, of volume v and temperature t, containing a certain quantity m_1 of hydrogen, a quantity m_2 of oxygen, and a quantity m_3 of water, one can ask what will be the course of events, that is to say, the evolution, of this system. We know that a reversible transformation will occur in this system according to the schema

$$2H_2 + O_2 \Leftrightarrow 2H_2O \tag{1}$$

reaching a perfectly defined point. But its definition is arrived at in terms of a certain function of the *entire* system, namely, its thermodynamic potential. It would be vain to attempt to indicate the direction of evolution of the system by speaking of only one or the other of its components, such as H_2O for example. And yet it is by studies concentrating on a species considered separately that biologists have for the most part sought to account for the direction of evolution in the organic world. We should profit instead from the physical chemistry model: how are we to conceive the march of evolution in the organic world?

We confront a system, vast it is true, but nonetheless of finite dimensions. That system is comprised of certain quantities of diverse species of living organisms, as well as inanimate components such as the oxygen in the air, etc. The state of this system at a given instant t will be defined by indicating the quantities $X_1, X_2, \ldots$ of the various components, as well as the values of certain parameters P serving to complete the description of the system, as for example the climate, topography, etc. In addition, to complete the tableau, it will usually be necessary to know certain other parameters Q, serving to define the character of the component species, that character being itself susceptible to alteration over time.

Now the changes that will take place in the composition of the living part of the system will fall into two categories: there will be, first, changes in the distribution of the matter in the system among the quantities $X_1, X_2, \ldots$ of the various species of biological organisms and the other components of the system. These changes constitute what we will call the *inter-species* evolution of the system.

Alongside these inter-species alterations there will be others, constituting a second category of changes, which will occur within each species itself. These alterations could be translated as alterations in the parameters Q by which the descriptions of the species are effectuated. However, we will be more consistent if we interpret the changes in the second category, as well as those of the first, as changes in the distribution of matter in the system, but this time among the components of each species. For it is evident that a species is not entirely homogeneous, but consists, on the contrary, of an aggregation of types that range in a more or less continuous manner between the extreme limits of that species. In fact, a complete description of a species would require an indication of the law of distribution (the frequency) of its various characteristics. For example, for the human species, a complete description would require the specification of a function $f(h)$ such that the quantity $f(h)\,dh$ gives us the proportion of all men whose height is comprised between the limits h and $h + dh$, let us say between 1.30 and 1.31 meters, between 1.31 and 1.32 meters, etc., and similarly for all the other characteristics of the species. The changes in the second category will thus translate as alterations in the parameters that characterize the distribution functions f. We will call these *intra-species* alterations, and their totality will be the *intra-species* evolution of the system.

Now there is a great contrast in the rapidity to which inter-species and intra-species evolution, respectively, are susceptible. The first can move in steps that are excessively swift, as in the case of a natural catastrophe which in a stroke annihilates an entire population of biological organisms or in the case of the invasion of virgin terrain by a species foreign to the territory up to that point, with more or less complete destruction of an indigenous species.

Intra-species evolution, by contrast, is always or nearly always effectuated by changes that are either very gradual or very rare. This evolution is thus characterized by its very slow march in comparison to that of inter-species evolution. In fact, in the discussion of the latter it will be permissible to neglect intra-species changes, changes in the parameters Q or the functions f, whenever the time that passes during the period of observation is brief in comparison to the time necessary for notable intra-species changes. From a practical standpoint, a very convenient situation arises from this circumstance. We will often be able to treat inter-species and intra-species evolution as two isolated subjects, although, in reality, the two phenomena take place simultaneously and mutually influence each other.

This is, then, our quantitative conception of the problem of evolution in the living world: we no longer ask simply what are the organisms which survive, but what is the quantity of each species and of each of its component types that is found living in the system at a given instant. In other words, what is the biological composition of the given system, the distribution of matter among its components, at a given instant; in addition, what changes will occur in this distribution? Do these changes tend toward a specific state, and if so, what are the characteristics of that state?

It is not yet possible to indicate in precise terms the law of evolution of the organic system, but we may be convinced that once formulated, this law will be given expression by saying that a certain characteristic function of the system as an *entire* system varies in a defined way, that it tends for example toward a maximum.

We have noted a certain analogy between a system in biological evolution on the one part, and a system in physicochemical transformation on the other. We must not suppose that this is a superficial analogy, from which we will extract robust conclusions by a "reasoning" lacking in reason. It is not an accidental analogy that interests us here, but the fundamental fact that we are occupied with a problem whose analysis is identical in its general form with the analysis of the problem of a system in physicochemical transformation. In one case as in the other we seek the laws that determine the distribution of matter between the components of the system. If the components are defined in a different manner in the two cases, that will doubtlessly greatly influence the methods we must follow in developing our analysis, but it will not in any way change the fact that the base from which we start will be identical in the two cases.

Several simple formulas will be useful for clarifying the exposition. We will say very generally that the rate of increase of the quantity X_i of a component S_i will depend, at each instant, on the quantities $X_1, X_2, \ldots$ of all the components, as well as the parameters Q which define the character of each component, and the parameters P which serve to complete the definition of the state of the system. If then, in the usual notation, we represent by the symbol dX_i/dt the rate of increase of the quantity X_i, we can put

$$\frac{dX_i}{dt} = F_i(X_1, X_2, \ldots, X_n, P, Q) \tag{2}$$

This equation is nothing other than the expression, in mathematical symbols, of the fact that the rate dX_i/dt is determined by (is a *function* of) the quantities written in parentheses; in addition, for each X the function F takes, in general, a specific form.

Now in this general formula the analytical expression applies equally to the case of a system in physicochemical transformation and to the case of a system within which biological evolution unfolds. It is only when we assign the functions F their specific forms, appropriate for the case being considered, that the analysis diverges for the two problems. In the case of a physicochemical transformation, equation (2) will take, for example, the form of the law of motion of masses. In the case of biological evolution we must seek entirely novel forms for the functions F. However, the fundamental formula of the analysis is the same in both cases, and the identity in form between the two problems extends as well to other important aspects that it is useful to examine, not without noting certain specific and very characteristic differences that superimpose themselves on the identity of the fundamental principles.

Linking equation. The changes in the distribution of matter in a system among its various components are accomplished by what we call *transformations* of matter. For example, in a system comprising a certain quantity (mass) of hydrogen, H_2, another of oxygen, O_2, and a third of water, H_2O, the distribution of the total matter of the system among its three components is susceptible to alteration by the fact that a transformation takes place according to the schema

$$2H_2 + O_2 \Leftrightarrow 2H_2O \quad (1)$$

This schema gives us no information as to whether the transformation will actually take place. All that it tells us is that, if it takes place, it can only proceed in conformity with the schema. One could thus say that the schema separates the transformations that are possible from all those that are impossible. Water, for example, would not be formed by a physicochemical transformation in a system entirely devoid of hydrogen.

The schema, without telling us which transformations will actually take place, thus poses certain conditions which every physicochemical transformation must satisfy to be realized. An equation, such as schema (1), which imposes certain conditions on possible transformations, which expresses certain linkages between the quantities of the various components taking part in the transformation, will be called a *linking equation*.

Linkages of systems in biological evolution. Our example of a linking equation has been taken from physical chemistry. What will be the character of the linkages within a system in biological evolution?

It is quite clear that certain transformations are possible, and others impossible, in our biological system. Grass transforms itself into cattle, or into sheep, etc., according to certain more or less familiar mechanical and physiological processes. Nowadays grass never transforms itself into mammoths, although at a certain period in the history of our world this transformation, or some other very analogous one, had not only been a possible event, but a daily occurrence.

In the organic world, a source of linkages which limit the possible transformations thus resides in a characteristic property of living matter: the formation of a new quantity of matter of a given species can only be effectuated in the presence of a pre-formed quantity of the same species. Given this presence, the formation of a new quantity of living matter can occur in two distinct manners. The new quantity can come to be added to the body of an individual which, itself, essentially conserves its identity and its characteristic appearance. That is what occurs every time we digest our dinner.

But in certain circumstances the process of assimilation takes a very singular direction. In place of simply adding to the body of the individual while preserving its identity and shape, the new substance comes to provide the materials necessary for the growth of a special part of the parent individual, this part developing more or less separately as a copy of the parent individual. The details of this process are well enough known, and for the purpose that we propose here it is not necessary to recall them. It suffices to emphasize that the result is a copy, which resembles the parent, but is not identical to it. Here we see an important difference between linkages in biological evolution and linkages in physicochemical evolution. The latter have a rigid character, inalterable, independent of time. The linkages in biological evolution, by constrast, have a certain elasticity. An eaglet never hatches from a duck egg, it is true. But a white duck may very well have bicolor ducklings. The character of the infant is not absolutely defined by that of its parents. It is true that certain extreme departures between the parents and their immediate descendants are precluded; that is why we stated earlier that the formation of a new quantity of living matter is possible only in the presence of a pre-formed quantity of the same species. But within certain limits the descendants of the same parents vary according to laws of heredity of which our knowledge is today still fragmentary. Whatever they may be, it is certain that the linkages in biological evolution are represented, not by simple equations similar to schema (1), but by *frequencies*, that is, by certain functions that indicate the probability that a given difference would be observed between the characteristics of the parents and those of their descendants. And, although our knowledge of these continuous or discontinuous functions is still very imperfect, we know from the observations of paleontologists that these functions are of a form that permits the linkages to change slowly with time. That is why a transformation from grass (or some analogous substance) into mammoths, which was very possible formerly, now no longer is, while certain other transformations, which occur everywhere today, would have been impossible in an earlier geological era.

We can, in this order of ideas, expand upon what was said earlier about a certain limitation in the works of biometrists of the conventional school. We can now say that these authors have been preoccupied almost exclusively with the *linkages* to which organic evolution is subject.

The situation is analogous to that which existed in chemical science when it was still preoccupied entirely with the study of schemas of reaction such as:

$$2H_2 + O_2 = 2H_2O \tag{3}$$

without posing the question, to what point the reaction would actually progress under given conditions, and, what is more, without seeking the laws that determine this point of equilibrium.

An important consequence results from the fact that the linkages in biological evolution are functions of time, while those in physicochemical evolution are fixed. Physicochemical evolution is a process that *terminates*. Biological evolution, by contrast, is almost certainly a process *without end*. From the state $2H_2 + O_2$ to the state $2H_2O$ there exists, so to speak, a sole path by which one can come and go from one extremity to the other, in a straight line. There thus exist, in this case, two limits, between which all the possible states are contained.

It is entirely different for systems in biological evolution. While the differentiation of elements and chemical compounds is a differentiation in *substance*, the differentiation of living organisms is essentially a differentiation in *structure*. Now, the differentiations in substance which are possible in the framework of chemical elements and compounds, while very numerous, are none the less definitively limited. The differentiations in structure, on the contrary, are not in principle subject to any limitation. We must therefore suppose that as long as the conditions that render life possible on our globe exist, biological evolution will continue without end. If a limit is imposed one day, it will not be due to the intrinsic character of evolution, but to cosmic changes that succeed in extinguishing the last sparks of terrestrial life.

Until then, since intra-species changes are very slow compared to inter-species changes, we can understand that biological systems will often be found in a quasi-stationary state, the inter-species changes having nearly reached their termination point, corresponding to the instantaneous character of the species. This termination point itself changes slowly to the extent that intra-species evolution progresses. Other quasi-stationary states* will be produced as well, when a slow

* Such states have sometimes been called "moving equilibria." However, we are not dealing here with true equilibria.

inter-species change gives rise to a series of other inter-species changes, which are themselves capable of much more rapid adjustment.

From what has preceded we conceive the evolution of a material system as a progressive change in the distribution of its matter among its diverse components. It is also this concept that will form the basis of our analytical treatment of the phenomenon. This essentially passive manner of regarding the phenomenon would be entirely insufficient for its critical study, however. The actual distribution, and the alterations which take place in it, are assuredly not passive states or effects. On the contrary, they are accompanied by certain energy processes of extreme importance. Every biological organism is, in effect, a transformer of energy, possessing certain properties of an interest and importance of the first magnitude. The study of the phenomena which a biological association presents should thus include certain chapters dedicated to the analysis of energy relations in such an association.

Here again it is useful to note certain analogies, as well as certain contrasts, between the energy data that we possess for physicochemical systems on the one hand and biological systems on the other. In the study of the first it is ensembles of molecules that present themselves easily to our direct observation, and in consequence, to our analysis. The individuals themselves, that is to say, the molecules and atoms, reveal themselves only indirectly or by the most refined observations. Hence, the principles of energetics applicable to these systems have been formulated for the most part directly in terms of the properties of ensembles. In the case of biological systems that order is inverted. The individuals generally present themselves to our observation and analysis without great difficulty. It is when we seek to account for the collective effects of the actions of individuals that we encounter a problem whose solution will require all the resources of our scientific spirit. And yet it is in the discussion of these collective effects that we should most probably seek the expression of the law of organic evolution, just as the law of physicochemical evolution is most easily expressed in terms of the collective effects of the molecules taking part in the transformations to which the system is susceptible.

CHAPTER 2

The term "evolution" is so solidly established both in everyday language and in scientific jargon that we accept it habitually, without sensing the need to analyze its exact meaning. We have used it in that way in what has preceded, recognizing that the reader will know sufficiently well what is intended.

However, the concept of evolution is assuredly not among those that must be accepted as impossible to analyze and define by the aid of other, more basic concepts. It would be treating our subject very superficially to fail to establish the fundamental ideas as clearly as possible, and to bring out certain quite sensitive difficulties that touch on the very basis of all of our discussions.

If we examine what our thoughts are when we speak of the evolution of a given system, we find first of all that the evolution of a system is its history, that is to say, the succession of diverse states through which it passes over time. However, this description of what the word evolution signifies for us evidently omits something essential. One would hardly call evolution the history of a system whose movement was purely periodic, infinitely repeating its journey through the same orbit. We said above that evolution consisted in *progressive* changes of the system. What then does this word progressive mean? The question is very delicate. Bertrand Russell, with his characteristic humor, offered the comment that "a process which led from amoeba to man appeared to the philosophers to be obviously a progress — though whether the amoeba would agree with this opinion is not known." To speak seriously, the idea of progress is based essentially on subjective sentiments, as Herbert Spencer has already recognized. We will shortly examine the objective foundations of the idea of progress, if in truth such foundations exist. But let us again note in passing that, even from a naive point of view, it is not always true that the evolution of biological organisms has taken place in a direction that we would call progress. Parasitic species are often adapted to their overly comfortable milieu by modifications which to us appear decidedly retrograde.

But there is a much more fundamental reason to reject a facile definition of evolution in terms of "progress." The idea of progress itself implies another more basic idea, namely, the idea of the passage of time in a certain uniquely characterized direction, the direction from the past to the future. When we seek an objective basis for this characterization, we find ourselves confronting unexpected and bizarre difficulties.

In classical mechanics a certain variable t, which it is convenient to identify with time, enters in calculations in such a manner that nothing is changed in the laws of motion if in place of t we write $-t$. From this perspective, mechanics thus does not distinguish between the past and the future. However, it would be more correct to say that time, in the way that we ordinarily comprehend it, does not enter at all in classical mechanics, and that it is only through a pedagogical error that we continue to teach this subject as if time played an essential role therein. For example, it is a fairly transparent error to develop the laws of motion with the aid of Atwood's machine,* making use of a clock or a metronome for marking "time." For we have the following alternatives:

Possibly we do not explain the action of the clock, which is itself a mechanism. The student then senses vaguely that there is something missing in our exposition; what results is confusion that has nothing to do with the subject about which we had intended to inform him.

Or possibly we say that the mechanism of the clock — a pendulum clock for example — will be explained later, when we arrive at the necessary point in the development of the subject. We then fall into a vicious circle; for we at first attempt to develop the laws of motion with the aid of a certain mechanism, then, later, we attempt to explain this mechanism by the laws developed with its aid.

The fact is that the clock is not necessary at all, and what is even worse, it obscures the true nature of the laws that are to be developed, and which, in reality, do not contain time as we ordinarily understand it.

The device to employ would consist of *two* independent Atwood machines, of which the first would be installed with the two suspended masses equal, so that its movement would be effectuated under the re-

---

* *Translators' note*: George Atwood's (1746–1807) machine consisted of two unequal masses connected by a silk thread that passed over a pulley. Atwood used the time required for the heavier mass to drop a fixed distance when released to estimate its acceleration, which, from Newton's second law of motion, should be proportional to the difference between the weights (the motive force) divided by their combined mass. A metronome alongside the machine served as a counter.

gime of a resultant force of zero, while the other machine would be handled in the usual way in developing the laws of motion. One would obtain equations resembling those to which we are accustomed, but in place of the symbol t, which we would understand to represent time, we would find a symbol x, which was nothing other than the *distance* traversed by the masses of the first Atwood machine. The whole system of mechanics which is based on fundamental laws would be recognized as a system of relations between the *positions* of certain points. One would then see clearly that no usage had been made of the idea of the passage of time.[*]

Actually we do possess a faculty which permits us to associate equal lapses of time more or less exactly with equal displacements of the masses of the first Atwood machine. But this observation belongs to psychology and not to physics. It is nonetheless very important, for it definitively coordinates with certain subjective data, that is, our impressions of the *length* of a lapse of time, objective data, that is, the lengths corresponding to the displacements of a certain mass. This is a service analogous to that which a prism renders when it coordinates with our subjective senses of violet, blue, green, yellow, etc., objective data, that is, certain wavelengths of light.

But when we seek an objective base for the direction from the past to the future, classical mechanics fails us. And, without entering into details, we note briefly that relativistic mechanics also does not give us satisfactory indications on this subject.

By contrast, it appears at first glance that thermodynamics furnishes an objective criterion that distinguishes the direction of the progress of time, and there is no shortage of authors to attribute our sense of this direction to the existence of *irreversible* processes in a thermodynamic sense.[†] Conforming to the law of the dissipation of energy (the *second*

---

[*] However, one would have to concede the following: 1) that we have the means, or the faculty, to distinguish the simultaneity of the positions of the various points that are of concern; 2) that we have the means or the faculty to recognize the identity of these points in their various positions; 3) that we have independent means to judge the equality of two masses. One such means is given in practice by the equal deformation of an elastic body, such as a spring balance.

[†] G. N. Lewis in his work, *The Anatomy of Science* (Yale University Press, 1926), p. 144, accorded to W. S. Franklin (1910) the merit of having been first to formulate this idea. That is a misunderstanding. The idea doubtlessly has a much earlier origin. W. Ostwald, in his book *Die Energie* (J. A. Barth, Leipzig, 1908), p. 86, attributes it to

law of thermodynamics), heat always passes from a warmer body to a colder body, and never in the opposite direction, so long as one is considering a simple transmission of heat by conduction. The same law establishes that when two vessels, each containing a particular gas, are brought into communication, the two gases mix, and the final state is such that each vessel contains a certain quantity of each of the two gases. Never, this law tells us, will one find the inverse process; two vessels brought into communication, each containing a mixture of the two gases, will not ultimately be found to each contain a single one of the two gases by a simple process of separation by diffusion. In other words, diffusion is always a process of mixing, never of separation.

Do we have here an objective criterion of the positive direction of the passage of time? It appears at first sight that we do, for we can say that the instant at which the two gases are mixed is later than the instant at which they are separated.

From a practical point of view it is certainly thus, and we did not hesitate earlier in speaking of the second law of thermodynamics as the law of physicochemical evolution. But, in principle, this criterion also fails us, because the theory tells us that it is highly improbable that the spontaneous separation of the two gases would occur, but that it is not impossible.

A model that can be developed easily will inform us on this point, and by an artifice, we can with its aid even positively demonstrate this miraculous and highly improbable separation, even capturing it at the moment it occurs.

We cannot follow the motion of molecules which mix or separate. But the principle will be the same if we substitute for them, for example, a game of cards or a quantity of numbered balls. Here also it appears at first sight that we will have a criterion for the direction of the passage of time. If we are given two lists of cards in the order they were found on two different occasions during the operation of shuffling, we would not hesitate to say that the list on which they appear all in order, each color separately, is the list of their anterior arrangement, and that the other list, on which the cards are mixed in no apparent order, is that of their ulterior arrangement. The passage of time translates as the passage of the

---

Descoudres. It was well known to all of those who were present at Ostwald's lectures; he also mentions it in his *Vorlesungen über Naturphilosophie*, Leipzig, 1902, p. 275.

cards from an improbable arrangement (all in order) to a more probable arrangement (mixed).

Have we truly found an objective criterion for the direction of the passage of time? I can show that we have not, that our criterion is an illusion. In place of a game of cards we will make use of a quantity of black and white balls. Fifty black balls are found in urn A, and fifty white balls in urn B. We draw at random one ball from urn A and another from urn B, and we return to urn B the ball coming from urn A, and vice versa. We mix the contents of both urns well, and repeat the operation of taking one ball at random from each urn and returning it to the opposite urn. After each draft we note the contents of the two urns. The diagram in Figure 1 represents the number of black balls in urn A after successive drafts. As is completely natural, one sees that the number of these balls decreases fairly systematically at the beginning of the operations. However, at the thirteenth draft the descent of the curve is interrupted momentarily by the return of a black ball to urn A. We should obviously expect this now that urn B contains several black balls. Ultimately the curve essentially reaches a stationary level corresponding to 25 black balls, half of the initial contents. But around this mean we observe both negative and positive fluctuations; sometimes the number of black balls is less than 25 and sometimes it exceeds that number. Would it be possible that one of these fluctuations reaches the number 25 above the mean, so that all 50 black balls are again found in urn A from which they were taken initially?

One would then see all the black balls gradually reassemble in urn A, and our diagram would indicate this miraculous drawing by a curve which rose in the manner of a stairway, from the mean level of 25 to the initial level of 50. This miraculous drawing has actually been realized in the experience registered in the diagram in Figure 2.

In the preceding paragraph we began the recital of our experience by the statement that "fifty black balls *are found* in urn A." We have observed a discreet silence on the subject of the manner by which these fifty black balls had entered into urn A. The fact is that they were introduced in the course of one hundred drafts resembling the series already described, this latter simply forming the continuation of an earlier series. Only before commencing the first series, the balls were all numbered, from 1 to 100, and throughout both series a protocol was followed to record the numbers of the balls contained in the two urns. In this manner, it was possible to take account at each instant of the detailed con-

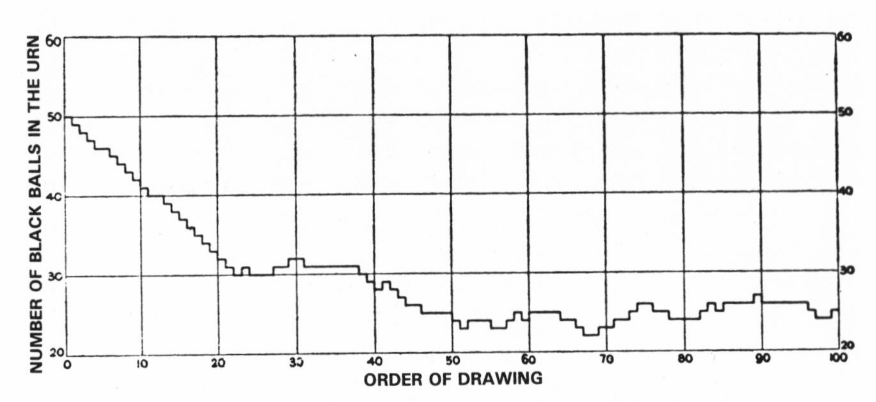

Figure 1. Curve representing the number of black balls in urn A after *n* drawings, beginning from a time zero taken as the hundredth draft.

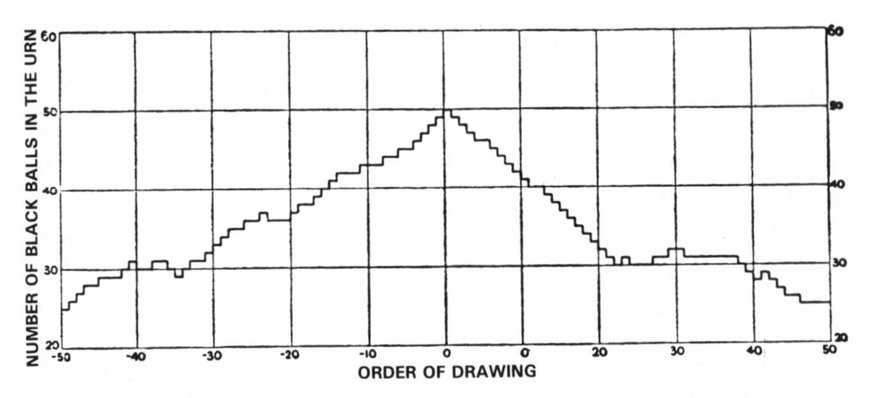

Figure 2. Curve representing the number of black balls in urn A after *n* drawings, from the fiftieth draft before time zero to the fiftieth draft after time zero.

tents of the urns. The balls in urn A were blackened *at the end of the hundredth draft*, that is to say, just before the commencement of the second series. But the protocol of recording the numbers of the balls also permits us to trace their peregrinations, and in that way to inform ourselves of the location of the 50 black balls at the outset of each draft, even before they were blackened. If we again construct the curve, this time representing the entire series, we see in Figure 2* that this curve begins in effect near the mean of 25 black balls and rises in a stepwise pattern until the moment that all of the black balls are found reassembled in urn A. We have thus actually witnessed this miraculous drawing, which has led to the "highly improbable" arrangement in which the black and white balls are found sorted and located separately, each in its own urn. The curve representing the succession of drafts, which appeared so distinctly asymmetrical when we began to trace it at the moment all the black balls were in urn A, has acquired an essentially symmetrical appearance to the left as well as the right of its summit, now that it has been completed so as to also reveal the events by which the black balls were assembled in A. On viewing the entire curve one could not say whether the events that it represents unfolded from left to right or the inverse. Our model has thus not furnished us an objective criterion for the direction of the passage of time after all.

It would require a strange mental aberration to take from us the faculty of distinguishing the past from the future, and it is difficult, if not impossible, to make sense of the bizarre confusion that would result for us from the loss of our subjective sense of time. But our model or its graphic representation permits us to place a person who has not witnessed the drafts in an analogous situation, that is to say, in the necessity of judging the chronological order of two instants by the so-called objective criterion, the contents of the two urns at these two instants. If, for example, we indicate to him that at a certain instant, during the series of random drafts, urn A contained 50 black balls, and at another 30, he would judge, from a naive point of view, that the 50 balls corresponded to an instant anterior to that of the 30 balls. We, who have witnessed the drafts, know that before as well as after the 50 black balls were assembled in urn A there was a moment when 30 black balls were found there, and that it is impossible to judge, solely on the basis of the numbers cited, which was in fact the anterior instant. If, on the other hand, we reveal to the person who must make the judgment that the balls were not blackened until *after* the hundredth draft, then his judgment is no longer objective, for it will be based on an indication "after" furnished by our

* To adapt the diagram to the format of the book, the curve in Figure 2 has only been traced from the fiftieth to the hundred fiftieth drawing, the time zero having been placed at the hundredth drawing.

own subjective judgment. In sum, our model is incapable of giving an objective criterion of the positive direction of time.

It is the same for the indications that the phenomena called irreversible in thermodynamics should give us as to the direction of time. These phenomena are essentially of the nature of mixings. The illusion that they indicate to us the positive direction of the march of time is due to the fact that we have commenced our observations at an instant when the system on which we are operating is found in a state very far from the mean (the equilibrium), and to the fact that fluctuations large enough to return it nearly to its initial state are rare to the point of never being observed during a time commensurate with the human lifetime or even the lifetime of our entire race.

As for the very small fluctuations on both sides of the mean in our model, they in fact have their representation in thermal phenomena. The mechanical theory of heat, as well as experimental physics, is familiar with these fluctuations, these small departures from the mean state, which reveal themselves to observation under certain very special conditions. We will see shortly what importance these fluctuations, ordinarily on an ultra-microscopic scale, can have for the philosophy of living matter and the problem of free will.

We have seen that a process of mixing is not capable of furnishing us an objective criterion for the direction of the passage of time and that it is essentially for this reason that, in principle, the second law of thermodynamics also cannot. At present there thus remains a certain mystery respecting the very foundations of every discussion of the phenomenon of evolution. Is the impression we have of a specific direction to the passage of time a purely subjective illusion? If it were, what sense could it give to the idea of a progressive evolution? The answer to this question escapes us today. It is without doubt intimately tied to the classical problem of causality.

CHAPTER 3

The model with which we are occupied is capable of illustrating still other principles which are important for our study. In a certain sense the operation of random drawings is *irreversible.* It would be vain to continue to make drafts in the fashion indicated, in hopes of one day seeing all of the black balls again reassemble in urn A, where they were found after the hundredth draft. Not that this event is impossible, but it is of such a high improbability that it would be senseless to expect to see it realized in a practicable length of time. We know well that if we are permitted to look into the urns and select the balls we withdraw with the aid of sight, we can very easily reassemble all the black balls in urn A. The operation of "drawing" is thus irreversible or reversible according to the means we are permitted to employ in choosing the balls that we extract from the urns. It is because we lack the means to operate on individual molecules that the process of molecular diffusion is irreversible in thermodynamics. But in their operations on masses of macroscopic size, biological organisms also frequently find themselves facing a type of irreversibility practically as insurmountable as that of thermodynamics. If a substance of great economic value, such as gold, for example, is found in some location in a state of extreme dispersion, mixed with valueless materials, its dispersion is for us practically an irreversible process, in the sense that it cannot be reversed (so as to concentrate the ore) without employing means excluded for reasons of economy. Now the inorganic forces of nature are in large part forces tending toward dispersion. Biological organisms, by contrast, depend for their existence on sources of matter and energy in a somewhat concentrated state. For this reason they possess more or less perfected organs and faculties, by which they effectuate the concentration of the materials they need. This concentration often takes the form of a "sorting" of valued materials which are found dispersed among other materials or objects without value for the organism. Whether this dispersion is irreversible or reversible for a given species will thus depend on the degree of perfection of its organs and its faculties of sorting. In addition, we see that in this case it no longer suffices to distinguish in an absolute manner between reversibility and irreversibility; it is necessary to envisage different degrees of irreversibility, corresponding to different degrees of perfection of the organs and faculties of sorting among different biological species. A dispersion that is irreversible for an organism of low intelligence, which depends almost entirely on chance encounters with the materials it wants, can be reversible (that is, quite capable of reconcentration) for an organism, such as man, which clearly recognizes the situation and possesses more or less refined means to take advantage of it.

It is precisely his wealth in faculties of sorting which has given man his dominant position and his worldwide distribution. Our entire system of agriculture, mines, manufactures and transport is nothing other than a vast device to assure to each a sufficient concentration, in his immediate vicinity, of the materials and energy that are indispensable or advantageous to him.

Here we have an indication of the direction in which the energetics of systems in organic evolution should develop. Like ourselves of the human race, all biological organisms are transformers of energy possessing properties that permit them to direct the energy they collect in ways serving to maintain them in their living state or to augment them in that state. From the point of view of physics, we thus see in organic evolution a process of distribution and redistribution of the matter belonging to the system among a number of transformers of this type. This distribution will thus depend on the degree of perfection of the devices which permit each species of transformers to collect energy and to direct it in ways advantageous to its existence. At the base of the ladder of life we find those which still depend greatly on chance to bring them the energy and materials necessary to them. To the degree that we ascend the ladder of life, the probability of obtaining that which is necessary is rendered greater and greater and the organism surmounts simple chance by its more and more perfected sorting devices, such as touch, smell, and above all sight, the servants of faculties of a relatively developed intelligence. The energetics of organic evolution should thus examine the distribution and redistribution of matter in a system comprising of such transformers, in relation to the properties that permit each of these species of transformers to direct energy in the ways appropriate to guarantee its continuing existence or its growth. It is in this direction that the mathematical analysis of the problem ought to be developed. It is likely that this analysis will profit from an artifice that might seem bizarre. It is not necessary that the transformers on which our discussion is based be biological organisms that actually exist. It is only necessary that they have the characteristic properties that a biological organism can possess. To reduce the problem to its fundamental elements it will probably be advantageous to concern ourselves with idealized transformers much simpler than those we encounter in nature. This will be a process of abstraction, similar to those which have shown themselves so fertile in results in the other branches of science.

It is certainly the case that the biological organism is capable of reversing processes "irreversible in a macroscopic sense." Is it also possible for it to do the same for irreversible processes on a molecular scale, those "irreversible in a thermodynamic sense?" That is a much debated question,[*] a question not without considerable import in the philosophy of science.

From the earliest days in the history of the thermodynamic concept of irreversibility we find allusions to this problem. Sir William Thomson (Lord Kelvin) formulated in 1852 the principle of the second law[†] with these words: "It is impossible, by means of *inanimate* material agency, to derive mechanical effect from any portion of matter by cooling it below the temperature of the coldest of the surrounding objects."[‡] Since that first suggestion by Kelvin, a number of authors have more or less definitively taken up the question of whether biological organisms are above the reach of the second law. In this discussion it is necessary to guard against an error. It is true that the action of biological organisms often has the character of introducing order where disorder reigned, while inorganic actions generally tend to destroy order and to replace it by chaos, as should happen in conformity with the law of dissipation. But this establishment of order by living agents in no way demands an infraction of the law. As long as the earth receives an energy flux from the sun in a useable form (that is, a level above that of terrestrial energy) it is possible, without violating the second law, for certain parts of the terrestrial system to undergo changes tending to establish order, or organization.

---

[*] See for example the correspondence between Sir James Jeans and F. G. Donnan and E. A. Guggenheim, in *Nature*, 1934, pp. 99, 174, 530, 612, 869, 986.

[†] This law can be expressed in various forms, which are, however, equivalent.

[‡] *Translators' note*: Emphasis added by Lotka.

We are thus not forced to invoke an extraordinary faculty for biological organisms, which positions them to mock the law of dissipation; all the same, it remains possible that they possess such a faculty. It must be said that, if they possess it, they appear not to make use of it except in a manner and measure that is extremely limited. In any event, we cannot entirely neglect this possibility, owing to certain important consequences it would bring with it. For here we are approaching very near to the classical problem of free will.

Intuition imposes on us an irresistible conviction that we are the *cause* of events that we have successfully *willed*. It thus lends to our will an important role in the determination of the actual course of events. That conviction of a certain caprice in the control of the world does not accord well with the very general impression the classical studies in the physical sciences have given us.

The laws of physics do not seem to leave us any liberty. A certain present state of the world would completely determine its history to infinity, and this without recourse, in our formulas, to any term representing the intervention of *will*.

However, this perspective is not entirely logical. For to say that the formulas are sufficient to describe the course of events *in the way we observe them*, does not exclude the possibility that they are incapable of describing the *entire* true phenomenon, which one could imagine, includes important elements that escape our observation. To say that elements which systematically escape our observation cannot have any practical importance is to speak without reflection. Human thoughts, emotions, and aspirations are certainly not without practical importance, although it is wholly impossible for us to observe them, or even to positively establish their existence. It is the philosopher, not the practical man, who dares to doubt these things.

Recent developments in the physical sciences have more widely opened the door of doubt that our formulas completely describe the fundamental phenomena of the world. Not only have these modern developments cast our attention on the fact that our crude observations are only observations of global, mean, or statistical effects; but they have further demonstrated that all of our means of observation, all of our experimental devices, are by their very nature, and in principle, incapable of informing us in an exact and complete manner about the ultimate and fundamental data of physical systems. The famous "uncertainty principle" tells us that *no material device* permits us to obtain an exact measure of the position and at the same time the speed of a speck of matter. One or the other of these quantities can be observed with a precision that can be augmented without limit as our instruments are perfected. But one of the two being thus determined, the other can only be known approximately, and that not because of the imperfection of our instruments or experiments, but in virtue of a basic and intrinsic quality of things.

From a practical point of view the uncertainty principle is ordinarily of little importance as long as we are dealing with matter of large dimensions. It is only when our experiences or our reflections touch on phenomena at an atomic or sub-atomic scale that it no longer suffices to consider mean effects. We then find ourselves forced to take into account the deviations that individual particles undergo in their minor fluctuations around the mean allowed by the so-called thermodynamic *equilibrium*. There also the uncertainty principle renders the exact prediction of events impossible, and introduces something that appears to us capricious in the affairs of the world.

Now, as R. S. Lillie has pointed out, it is precisely a characteristic of biological organisms that their action depends on their fine structure, on microscopic and ultra-microscopic details of their bodies. Certain effects that unfold on a microscopic scale or finer are translated, by an organization designed precisely for that end, into macroscopic effects, such as the movement of its members by which the organism reacts on its environment. It follows that the element of caprice that ordinarily is only felt in atomic or sub-atomic events emerges, by the action of the organism, onto the sphere of phenomena observed by the naked eye. At the least we must be prepared to acknowledge these ideas as hypothetical possibilities.

From this perspective, what distinguishes organisms from inanimate matter is not the presence in the former and the absence in the latter of a specific quality, but the presence of a mechanism translating to a macroscopic scale certain effects which, in inanimate matter, cancel themselves out for our excessively gross observations.

There is not, in principle, anything mysterious in the emergence with which we are dealing here. It is very simply the emergence from regions which, by their minuscule scale, escape our observation, to the sphere of things knowable to our senses. I say this because certain authors have attached to the idea of emergence, of *emergent evolution*, a mystical significance, according to which there is in this idea something that characterizes certain relations and certain processes which are restricted exclusively to biological organisms. The emergence about which we have spoken is actually encountered each time we consider observations or relations that cross the barrier separating the visible from the invisible, that which can be sensed from that which cannot.* As for a mystical emergence, it introduces a new term into our discussion without introducing new facts. One must mistrust these verbalisms. We disguise our ignorance with words. We extinguish our curiosity without satisfying it. Nerve energy, vital force, and all similar terms, are names which introduce error into research on the subjects to which they attach themselves. It is not in this way that science advances. One must locate the subject and test the necessity of giving it a name before celebrating its baptism.† Is it not strange that a truth easily seized by the whimsical spirit of a Molière would have escaped the more serious thoughts of certain philosophers?

RECAPITULATION

Before considering more definitively the specific problems which will occupy us in subsequent chapters, it will be useful to briefly recapitulate the general principles we have noted thus far.

We have envisioned the evolution of a material system as the progressive redistribution of matter between its components. Depending on whether the latter are defined in physicochemical terms (elements, compounds, phases) or in biological terms (biological organisms and their habitat) we are dealing on the one hand with physicochemical evolution,

---

* There is nothing occult in the fact that Mr. Jourdain "emerges" from Cambon Street at the moment when I pass the corner of this street on my walk along Rivoli Street. Quite simply, the buildings have hidden the view from me.

† *Translators' note*: Lotka's comment in *Elements of Physical Biology* (p. 13) may be clearer: "If we have cause to hesitate in defining life, still more is it the part of wisdom to be very conservative in the coining and use of such phrases as *vital force*, *nerve energy*, and the like. Shall we not do well to follow the biblical example, and wait, to name the animal, until it is physically present to our senses?"

or on the other with organic evolution. In a system in physicochemical evolution, the linkages are of a fixed form, invariable with time. The study of systems in this category comprises on the one hand stoichiometry, which is concerned with relations between the quantities of matter (masses) taking part in physicochemical transformations; and on the other dynamics, or more exactly, the chemical energy balance, which involves relations between the transformations of energy that accompany the transformations of matter. In addition, we can be interested in general with the *rates* at which these transformations take place, or we can devote our attention more specifically to the case where the rates cancel, so that a stationary state is established. In other words, we can discuss the *kinetics* or the *statics* of systems in physicochemical evolution. Finally, in studying these systems, we can concern ourselves with molecular aggregates as a whole, as thermodynamics does in general, or we can push the analysis further and take into account events on a molecular, atomic, or even sub-atomic scale.

In contrast to this state of affairs, when the components of the system being considered are defined in biological terms, it is characteristic of the linkages that they are variable, functions of time. In consequence, we distinguish in this case two different aspects of the evolution of the system, namely: first, intra-species evolution, which is concerned with changes in the character of biological species, envisaged as a distribution of matter among the various types of which each species is composed; and second, inter-species evolution, which is concerned with changes in the distribution of the material of the system among the different biological species and their milieu. In addition, a treatment of the subject pretending to be absolutely complete should include a discussion: *a*) of relations between the quantities of matter (masses) entering into play, and *b*), relations between the transformations of energy which accompany the process of evolution. Here also, as for physicochemical systems, it will be entirely natural to divide the systematic discussion into two parts, one kinetic and one static, according to whether we are specifically concerned with the *rates* of change, or with the *stationary states* or quasi-stationary states toward which the system in evolution is tending, and which are sometimes established, more or less definitively, in such a system. Finally, we have stated that in physicochemical systems it is the global effects which lend themselves most easily to our observation, the individuals themselves (molecules, atoms, etc.) escaping our discernment; while, by contrast, in systems that include biological organisms the individuals are generally the objects of our direct observation, while global effects, the collective actions of those individuals, must be grasped by special methods.

We have noted certain difficulties that attach to the concept of *progressive* changes, not only because we lack a satisfactory objective definition of what constitutes progress, but also by reason of a fundamental uncertainty that exists about the direction of the passage of time. For this direction, we presently have no entirely satisfactory objective criterion. We are thus forced to develop our analysis as best we can, depending on our subjective judgment of the asymmetrical character of the passage of time. The variable t will enter in our discussions in the sense that we intuitively attach to the word "time." In this, we follow a custom that is ordinarily accepted without discussion in the analysis of natural phenomena. Still, it was necessary, in a systematic and rational study of systems in evolution, to highlight certain difficulties that touch on the very foundations of the subject.

A question that may be even more fundamental, which is introduced in the analysis of the problem of evolution, is born of the "uncertainty principle," according to which the very nature of physical phenomena renders it impossible for us to ever know the state of a physical system in enough detail to be able to deduce from this knowledge the ultimate course events will take. This uncertainty is felt particularly in phenomena on a molecular scale or finer. From the fact that the actions of biological organisms depend greatly on phenomena on this scale, we have drawn the provisional conclusion that free will could definitively range in the order of nature owing to the "uncertainty principle."

Having thus summarized concepts and fundamental principles, we are ready to enter into the analysis of the evolution of a biological system definitively; in other words, we are ready to develop the analytical theory of biological associations. We will begin the analysis with *biological stoichiometry*, that is, with a discussion of relations between the quantities of *matter* (the *masses*) contained in the various component species of the system. The discussion of the energy balance of such systems, of which we today possess only the rudiments, will form the subject of future chapters.*

---

* *Translators' note*: After finishing Part I of the *Théorie Analytique*, Lotka turned his attention to demography in Part 2, published five years later. The further chapters on

biological systems envisioned here and in the next chapter, which would have drawn on the *Elements of Physical Biology*, were not completed.

CHAPTER 4

Biological Stoichiometry

Recognizing that at every instant the rate of increase of each species in the system being studied depends on the quantity of that species and of every other species present, as well as the parameters P and Q, we have already noted that the analytical expression of this very general proposition takes the form:

$$
\left.
\begin{aligned}
\frac{dX_1}{dt} &= F_1(X_1,X_2, \ldots ,X_n;P,Q) \\
\frac{dX_2}{dt} &= F_2(X_1,X_2, \ldots ,X_n;P,Q) \\
&\cdots \\
\frac{dX_i}{dt} &= F_i(X_1,X_2, \ldots ,X_n;P,Q) \\
&\cdots \\
\frac{dX_n}{dt} &= F_n(X_1,X_2, \ldots ,X_n;P,Q)
\end{aligned}
\right\} \qquad (4)
$$

We have also noted that when we are considering a relatively short period of observation we can often neglect the changes in the parameters Q which express intra-species changes in form that occur very slowly.* We then address what we will call "pure inter-species evolution." In addition, we may be specifically interested in the special case in which the general conditions of the system (climate, topography, etc.) remain

* Alongside these *essential* changes in the basic character of the species, we will also have to consider certain more or less *adventitious* changes, that is, more or less superficial changes such as those in the age distribution of the individuals of the species. We will see later that in certain cases the effect of these changes reenters without difficulty into our analysis. It must be acknowledged that in other cases it introduces rather formidable complications. For the moment, we will neglect this factor.

constant. The parameters P are then also invariable, and we write more simply:

$$\left.\begin{aligned}\frac{dX_1}{dt} &= F_1(X_1, X_2, \ldots, X_n)\\ \frac{dX_2}{dt} &= F_2(X_1, X_2, \ldots, X_n)\\ &\cdots\\ \frac{dX_i}{dt} &= F_i(X_1, X_2, \ldots, X_n)\\ &\cdots\\ \frac{dX_n}{dt} &= F_n(X_1, X_2, \ldots, X_n)\end{aligned}\right\} \tag{5}$$

The precise character of the functions F would be examined through experimental observations. However, as we will see later, certain very general considerations permit us to extract conclusions of considerable weight from the fundamental equations.

Here our investigation touches on the special branch of biology called *ecology*. This science is concerned with assembling as many quantitative observations as possible on the interdependence among species and between species and their habitats. The methods employed to this end are now reasonably well established. For domestic species in their relation to the human race, simple agricultural statistics on current living capital, production, and consumption furnish the fundamental data. Similarly, when we consider experimental colonies of organisms of whatever type, observation presents very few difficulties. For wild species the problem is more difficult. Plants, which by their nature remain attached to a fixed place, also lend themselves relatively easily to a census, for which well established methods exist. Among animals, it is perhaps the aquatic species that offer themselves most conveniently to ecological inquiry. One has only to examine the contents of fish nets to obtain an initial indication of the relative frequencies of certain species. Refinements of this basic method, making use of fine nets and other special instruments, furnish quantitative indications on the flora and fauna of the waters themselves and of their margins. Dissection of the stomachs of fish and other aquatic animals gives information on the distribution of devoured species in the food of the devouring species. This method applies equally to terrestrial species and birds. Ecological studies of this type are of great economic importance, for the information they provide us on the utility or nuisance character of the animals belonging to our entourage. Work in this domain has also been actively pursued, so that a very considerable literature now exists on the subject.*

From a qualitative point of view the mutual dependence of biological species (which is translated in our analysis by the functions F) takes quite varied forms. This dependence can be of the *coordinate* or *subordinate* type, the first category comprising all of the cases in which two species find themselves in competition, either for a source of food or for another commodity common to them, for example the terrain that they inhabit. That competition can be passive, each of the species influencing the other indirectly by diminishing the food or some other commodity that offers itself to their common need. It can be active when the two species fight each other, in a combat that can be mortal, over the items from which both seek to profit. It goes without saying that competition of a mixed character, passive and more or less active, will occur in many cases.

Dependence of the *subordinate* type is that in which one of the two species is useful to the other. The most common case of this type is that in which a victim species, or a product of this species, serves as food for a second species. If the individuals of the victim species are themselves exploited by the predatory species, one distinguishes two cases: in the first, the victim remains alive, more or less damaged, while the predator feeds on the substance of the victim itself or on its incompletely digested products. The harm to the victim species may be relatively minor, as in the case of certain parasites that are irritating but not very injurious, or that of plants of which certain parts are consumed by herbivores. The harm can be grave, as in the case of parasites which weaken their host, sometimes very considerably, even to the point of menacing its life.

The second case of subordinate interdependence is that, very characteristic of carnivores, in which the victim is killed immediately by the predator, which devours it. Generally the victims are determined more or

---

* One will find very detailed information on this literature in the bibliographical notes in the work by R. N. Chapman, *Animal Ecology* (Burgess–Roseberry Co., Minneapolis, 1926, 1927).

less by chance, every edible individual which lets itself be caught being accepted by the predator. However, the human species has developed a very special system: he chooses his victims, not by chance but taking care to leave alive, either in the wild or in captivity, a "stock," which assures the continuing existence, in a sufficient quantity, of the victim species.

These different modes of mutual dependence between the species of biological organisms as well as several others are summarized in Table 1, which the reader can amplify as he pleases.

Linkages of species. In nature the relations between the various species become very complicated because the greater part eat a mixed diet containing a large number of other species; and because, for their part, many species have not one but numerous enemies which devour them. What results is a complex linkage, in which each form finds itself intertwined with a certain number of species to which it is subordinate, and a certain number of others which are subordinate to it. Relatively complete studies on this topic have so far been made only for a few groups of animals, such as the herring and the species which serve it directly or indirectly as food. Figure 3, taken from the research of A. C. Hardy, presents a good illustration of the complicated relations with which we are dealing.

If, in the general schema of linkages, we isolate a particular line, combining a first species with a second that devours it, a third which devours the second, and so forth, we obtain what is called a food chain. Very interesting reflections of a general character respecting these chains have been made by C. Elton. That author has pointed out that it is not pure chance which regulates the character and length of such chains. As Elton states, "there are very definite limits, both upper and lower, to the size of food which a carnivorous animal can eat. It cannot catch and destroy animals *above* a certain size, because it is not strong or skillful enough. ... At the same time a carnivore cannot subsist on animals *below* a certain size, because it becomes impossible at a certain point to catch enough in a given time to supply its needs." It follows from this that a food chain cannot comprise more than a certain number of members; that number rarely exceeds five.

One other remark is suggested by Table 1. The interaction between the various species is mutual, but it is not, in general, reciprocal. That is

Table 1. Mutual dependence of biological species

- Coordinate
  - Indirect: Competition (common food source)
  - Mixed: Both competition and mutual destruction
  - Direct: Mutual destruction without benefit to either species
- Subordinate
  - Bodies of one species are eaten by a second species
    - Encounters with the predator species are fatal for individuals of the prey species
      - Random destruction
      - Selective destruction
        - Without cultivation (game protected by hunting laws)
        - With cultivation (livestock)
    - Individuals of the prey species remain alive while the predator species feeds on them
      - The predator species passes its life on or in the body of the prey species
        - Parasitism
        - Physiological symbosis
      - The prey species leads a separate existence (herbivores)
      - The prey species is cultivated by the predator species (cereals)
  - Products of the bodies of one species are eaten by a second species
    - Seizure of products useful to the prey species
    - Utilization of wastes of another species

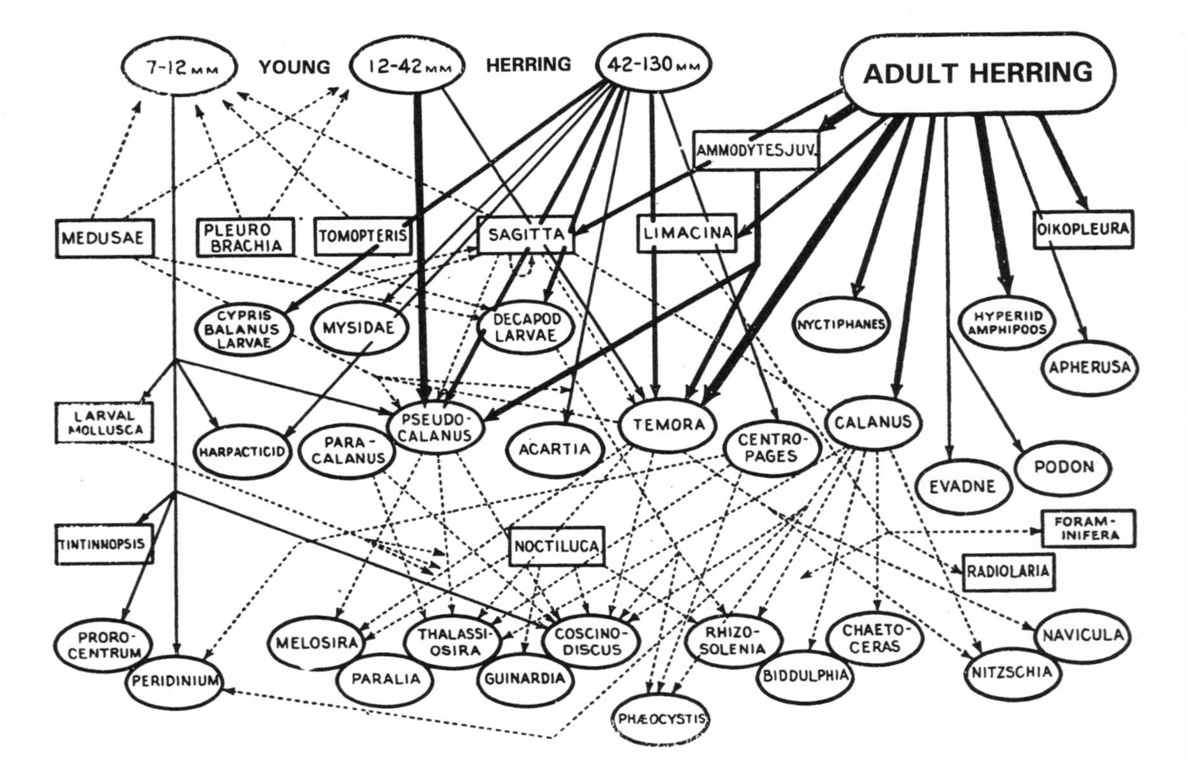

Figure 3. The herring and the species which serve it directly or indirectly as food. One sees the influence of the size of the young and adult fish by the choice of its food. (From A. C. Hardy, The herring in relation to its animate environment, Part I. Ministry of Agriculture and Fishery Investigations, 1924, Ser. 2, Vol. 7, No. 3. Reproduced by permission of the Controller of His Britannic Majesty's Stationery Office.)

to say (although there are notable exceptions), that one of the two species in a mutual relation is generally useful to the other, while the second is harmful to the first. The exceptions are the cases of symbiosis with reciprocal benefits. What interests us here, and which will interest us more a little later, in the chapters on biological dynamics, is that we encounter here, in the discussion of very general biological relations, a concept — the concept of positive and negative utility — which would appear at first sight essentially economic.

The fact is that our economic needs are born naturally of our biological needs, and that political economy relative to the human race is at its base only a very specialized political economy. The theory of value should seek its fundamental principles in biology, and in particular, in the analytical theory of biological associations. We will have occasion to return to this topic.

Having reviewed several of the most characteristic relations between biological organisms of various species, let us return to the analytical representation of these relations, the system of equations (5).

Let us look for the values of $X_1, X_2, \ldots, X_n$ which correspond to a stationary state, that is, the values for which the rates dX/dt all cancel. We thus pose:

$$F_1 = F_2 = F_3 = \ldots = F_n = 0 \tag{6}$$

and we obtain a set* of values:

$$X_1 = C_1, X_2 = C_2, \ldots, X_n = C_n \tag{7}$$

We introduce the new variables:

$$\left.\begin{aligned} x_1 &= X_1 - C_1 \\ x_2 &= X_2 - C_2 \\ &\cdots \\ x_n &= X_n - C_n \end{aligned}\right\} \tag{8}$$

* In general, there may exist none, one, or several sets of values satisfying equation (6) and corresponding to as many stationary states. In what follows we will fix our attention on a selection of these.

so that our equations (5) take the form:

$$\left.\begin{aligned}\frac{dx_1}{dt} &= f_1(x_1,x_2, \dots ,x_n)\\ \frac{dx_2}{dt} &= f_2(x_1,x_2, \dots ,x_n)\\ &\dots\\ \frac{dx_i}{dt} &= f_i(x_1,x_2, \dots ,x_n)\\ &\dots\\ \frac{dx_n}{dt} &= f_n(x_1,x_2, \dots ,x_n)\end{aligned}\right\} \tag{9}$$

Now, without knowing the precise form of the functions *f*, and being content to suppose that they can be developed in a Taylor series, it is possible to extract from our equations certain remarkable conclusions.

We thus pose, following Taylor:

$$\frac{dx_i}{dt} = a_{i1}x_1 + a_{i2}x_2 + \dots + a_{in}x_n + a_{i11}x_1^2 + a_{i12}x_1x_2 + \dots \tag{10}$$

assigning i all the values from 1 to n, so that we have a system of n simultaneous differential equations.

We know a *formal* solution of the system (9), which is of the form:

$$x_i = G_{i1}e^{\lambda_1 t} + G_{i2}e^{\lambda_2 t} + \dots + G_{i11}e^{2\lambda_1 t} + G_{i12}e^{(\lambda_1+\lambda_2)t} + \dots \tag{11}$$

that is, of the form of an ordered series in increasing powers $e^{\lambda_1 t}$, $e^{\lambda_2 t}$, ..., the coefficients λ being the roots of the characteristic equation:

$$\begin{vmatrix} a_{11}-\lambda & a_{12} & \cdots & a_{1n}\\ a_{21} & a_{22}-\lambda & \cdots & a_{2n}\\ \cdots & \cdots & \cdots & \cdots\\ a_{n1} & a_{n2} & \cdots & a_{nn}-\lambda \end{vmatrix} = D(\lambda) = 0 \tag{12}$$

When the series in (11) are convergent they represent the course of the increase or decrease in the quantities x (and in consequence X). But in any event, the roots λ give us important information about the character of the stationary state to which they relate. For in the immediate vicinity of that state, the x being sufficiently small, we can ignore nonlinear terms in the Taylor series (10) and in that case the solution (11) reduces to the linear terms:

$$x_i = G_{i1}e^{\lambda_1 t} + G_{i2}e^{\lambda_2 t} + \ldots \tag{13}$$

It is then clear that if all of the roots λ are negative, or complex with their real parts negative, we are dealing with a stable stationary state, for in that case, each x_i tends toward zero as t tends toward infinity.

In contrast and by the same reasoning, if only one of the roots λ is positive or has a real part that is positive, the stationary state cannot be stable.

More generally, the character of the stationary state depends on the roots λ. Even when the number of species, and in consequence the number of roots λ, is only two, a number of different combinations are possible, giving rise to as many distinct types of stationary states. The analytic criteria corresponding to these diverse types of stationary states having been discussed in detail in another volume in this series, by V. A. Kostitzin,* I will limit myself to reproducing here (Figure 4) a plate extracted from my *Elements of Physical Biology* (1925) which illustrates some of these types. The diagrams of this plate should be interpreted as follows:

To bring out clearly the characteristics of the different types of possible stationary states in the case of two biological species, it will help to eliminate the variable t from the equations:

$$\left.\begin{aligned} \frac{dx_1}{dt} &= f_1 \\ \frac{dx_2}{dt} &= f_2 \end{aligned}\right\} \tag{14}$$

---

* No. 96: Symbiose, Parasitisme et Évolution, 1934.

so that one has:

$$\frac{dx_1}{f_1} = \frac{dx_2}{f_2} \tag{15}$$

We represent the variables x_1, x_2 by the abscissas and ordinates in a rectangular system (Figure 4). Equation (15) then defines a family of integral curves all passing through the origin, that is to say, through the singular point:

$$f_1 = f_2 = 0 \tag{16}$$

which represents a stationary state. The shape of the integral curves in the neighborhood of the singular point depends on the roots of the characteristic equation (12). Beginning from an arbitrary initial state, the masses x_1, x_2 of the two biological species undergo alterations (increase or decrease), such that the point which represents the state of the system at each instant traces a curve on the diagram. We will say that the system traverses a path indicated by one or another of the "integral curves" in the illustration, the direction of movement being indicated by an arrow.

A selection of cases can be presented. When the roots λ_1 and λ_2 are both real and negative, the integral curves enter directly into the origin, the arrows being directed towards this point (Figure 4A). A similar case presents itself if the two roots are real and positive (Figure 4B), but here the stationary state is unstable, the arrows turning away from the origin. The case in Figure 4C is that in which the roots λ are both real, but one is positive and the other negative. The stationary state is unstable. The case of Figure 4D corresponds to two roots λ which are complex conjugates whose real part is negative. In this case the integral curves are spirals, circling the origin and approaching indefinitely near to it, without ever reaching it. This is evidently also a case of a stable stationary state, but the variables x_1, x_2 rather than approaching that state by a direct route, undergo damped oscillations and converge indefinitely toward that state without ever attaining it exactly. Case E is similar, but the origin represents an unstable state, the roots λ being complex with their real parts positive and the spirals being traced in a direction away from

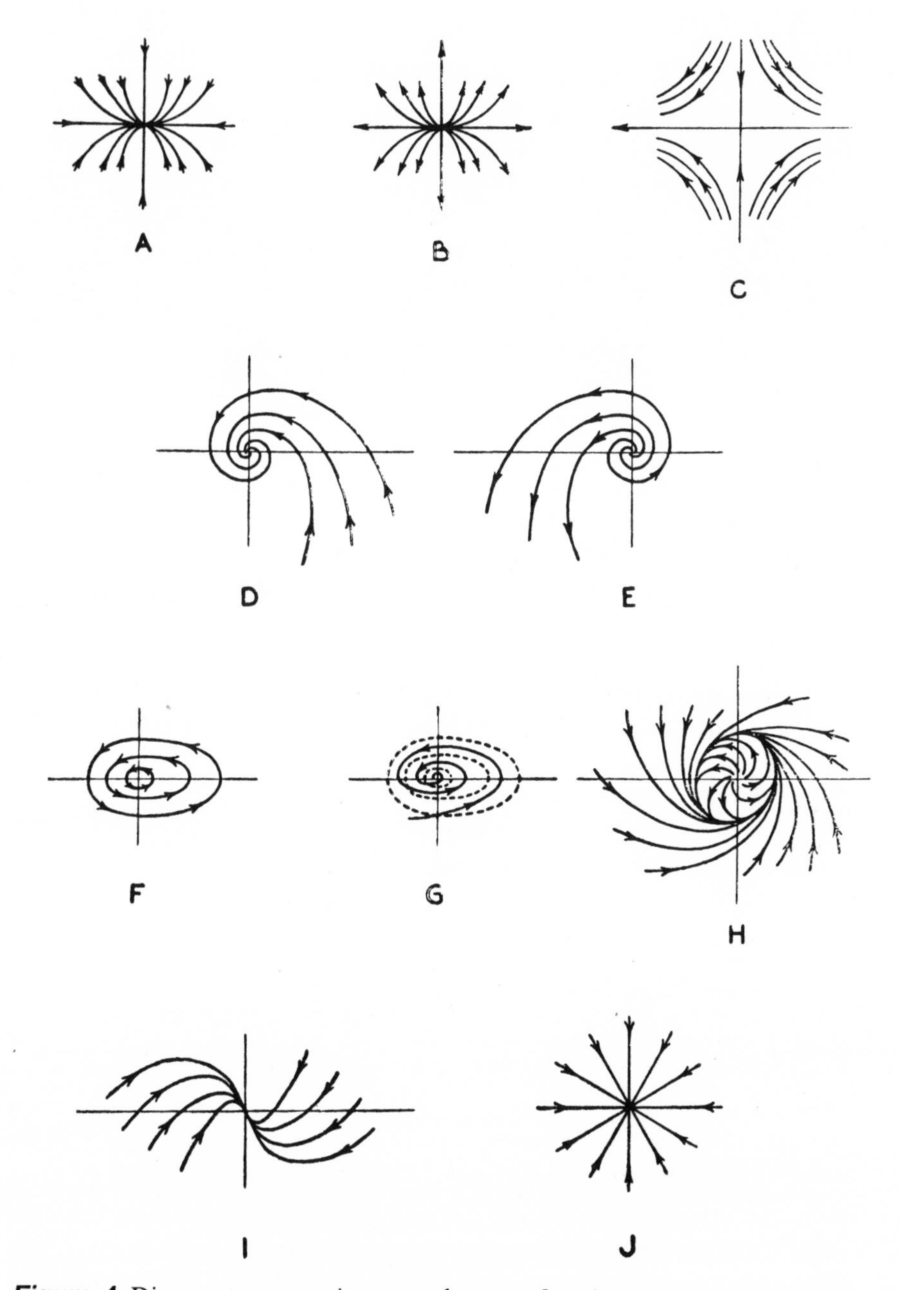

Figure 4. Diagrams representing several types of stationary states in a system comprising two biological species.

the origin.

Diagram F is encountered in certain cases when the roots λ are purely imaginary quantities. There are in that case periodic oscillations, not damped but continuing to infinity. That is an exceptional case, but of particular interest in theory if not in practice. More generally, the case where the two roots λ are purely imaginary quantities leads to spirals of type G or H.* Cases I and J present themselves when the characteristic equation has a double root.

Special Cases. In what has preceded, as in my earlier publications, the analysis has been developed from a general perspective† for the case of n species, n being any number. Many special cases resolve into the general case without any precautions other than to assign n its individual value and give other characteristics the form appropriate to the case in question. There are, however, exceptional cases which require treatment by special methods. They are those where certain of the coefficients a_{ij} in the Taylor series (10) cancel. Consideration of these cases, which are often of particular interest, must be reserved for a subsequent volume. Here, in order to conclude and to prepare for the Second Part of this study, we will again briefly note the case in which a single species develops in a manner apparently independent of its contemporaries. In fact, although in truth no species can really be independent, observation informs us that in certain circumstances the growth of a population of a given species follows a curve that can be entirely defined in terms of this species itself, without explicitly mentioning other species. Here the system of fundamental equations (5) reduces to a single one:

---

* For a more detailed discussion the reader is referred to the critical examination of the singular points of the integral curves, which has been developed particularly by H. Poincaré and E. Picard. An example of type G will be found in A. J. Lotka, "Contribution to quantitative parasitology," Journal of the Washington Academy of Sciences, 1923, p. 152.

† By a misunderstanding whose explanation escapes me, some authors (see for example, V. Volterra, *Leçons sur la Théorie Mathématique de la Lutte pour la Vie*, Gauthier–Villars, 1931, p. 4; R. Risser, *Applications de la Statistique et la Démographie à la Biologie*, Gauthier–Villars, 1932, p. ix) have spread the impression that I have only treated the case of two species. However, I have very clearly based the entire discussion on the general case of *n* species since my publications of 1912, and, in more detail, in my book *Elements of Physical Biology*, 1925.

$$\frac{dX}{dt} = F(X) \tag{17}$$

Let us assume $F(X)$ can be expanded in a Taylor series, and put:

$$\frac{dX}{dt} = a_0 + a_1 X + a_2 X^2 + a_3 X^3 + \ldots \tag{18}$$

As there must be at least one female in order for a population to be created, it is evident that the constant a_0 cancels.

The simplest case would then be:

$$\frac{dX}{dt} = a_1 X \tag{19}$$

$$X = X_0 e^{\alpha_1 t} \tag{20}$$

That is, growth according to the law of compound interest, called the "Malthusian" law. We know that under certain conditions populations increase, at least for a limited period, according to this law.

The Malthusian law evidently cannot continue to translate the growth of a population indefinitely, as the values it prescribes ultimately surpass all limits. There should in fact also be a value of X other than zero by which the rate of increase cancels. The expression dX/dt should thus have a factor $(C - X)$. The simplest expression which satisfies this condition, and at the same time cancels with X, is:

$$\frac{dX}{dt} = KX(C - X) \tag{21}$$

$$X = \frac{C}{1 + Qe^{-KCt}} \tag{22}$$

This is the law of Verhulst (1844) resuscitated independently by Pearl and Reed in 1920, and recognized by Lotka, in 1925, as a special case of his fundamental system (2); also recognized as a special case of the general expressions by Volterra in 1926.

There exists a fairly extensive literature on formulas appropriate to describe the growth of a population. A review will be found in the article by G. Teissier, "Croissance des populations et croissance des organismes: Examen historique et critique de quelques théories," *Annales de Physiologie et de Physiochimie Biologique*, 1928, p. 342. For our present purposes these indications must suffice. The case of one species considered by itself presents a field of study of such a vast reach that one or even several monographs will need to be dedicated to it. That comes about from the fact that there exists a certain species for which we possess an unrivaled abundance of information: the human species.

Analytical Theory of Biological Associations

"Dans les questions compliquées le bon sens a besoin d'être guidé par les résultats du calcul; les formules ne créent pas l'esprit de finesse, mais en facilitent l'usage." [In complicated questions good sense needs to be guided by the results of computation; the formulas do not create the sensitive mind but facilitate its use.]

Émile Borel

CHAPTER 1

Introduction

The species of living organisms are found in a mutual relation one with another, so that in truth it would be impossible to do a well rounded study of any single species without taking into account a considerable number of other species which influence it in one manner or another.

Nevertheless, there exist among the factors which are internal to a population of living organisms (such things as its births, deaths, and natural increase, etc.) a large number of relations that permit and even ask for special study, without it being necessary at every step to account explicitly for the influence of other species inhabiting the same locale. That study, in fact, constitutes a well defined body of research and findings, which we will encounter in the present monograph, where we consider specifically the human species, for which we possess an abundance of observational data.

Now, it must be pointed out that the demographic and sociological study of the human species involves a number of relations which, for other species, either do not exist at all or play a minor role. From that distinction a number of sociological and statistical problems derive that are peculiar to the human species, for example, the essentially monogamous laws of civilized peoples, and man's prolonged adolescence. Statistics on vital events, on sterility and fertility in marriage, on the family, orphans, etc., and the problems that arise from them, don't exist as it were except for our species, even though they are in essence particular aspects of the general phenomenon of biological reproduction. Although the application of certain of these "family" problems is entirely confined to a single species, the central position mankind occupies for us in the natural tableau gives these applications an importance of the first order. We must therefore be concerned with them.

Demographic Statistics. Without observational data it would be frivolous to develop a theoretical analysis of any natural situation. By contrast, as soon as observation has furnished us a quantity of empirical data, the problem arises — one would even say the necessity if we are to understand the data well — of analyzing the relations that exist between them. That analysis itself might still rest in an empirical framework, as for example in the purely arithmetical examination of the relation between variations in the density and death rate of a population. No one would deny that studies of this type are of great importance. They are essentially contained in the science we call *Demographic Statistics*.

However, while admitting the importance of such purely empirical studies, we will find more satisfying to the mind the more complete or at least deeper knowledge that is obtained when we have succeeded in taking into account not only empirical relations, for which the physical or logical reasons escape us, but also, and more particularly, the *necessary* relations (imposed by the laws of logic and physics) between the quantities serving to describe the state and changes in the state of a population.

The study of such necessary relations constitutes another body of science, which is called *Demographic Analysis*.*

Demographic Analysis. It might appear at first glance that the study of necessary relations would be a sterile occupation, for it can never reveal to us any fact whose truth is not already implied in the data on which the study is based. To this we answer that a truth which for our mind remains "implied" without being recognized is without practical value; more, that our mind is more demanding: it asks not only to un-

---

* *Translators' note*: The original text reads "L'étude de telles relations nécessaires constitue encore un corps de science, qu'on appelle l'*Analyse Démographique* («Population Analysis»)." We have followed Lotka's usual English usage in the translation.

derstand a body of facts, but also to recognize their linkage. It is precisely this linkage which forms the subject matter of demographic analysis; that is, the examination of general demography by deductive reasoning.*

Now the deductive method presupposes precise hypotheses, whether expressed definitively in words or contained implicitly in the terms of the propositions. To shelter us most completely from errors, it would be desirable that all of our hypotheses be explicitly declared. However, certain hypotheses are so evident that it would be superfluous to give them special emphasis.

It will be admitted, for example, that the annual increase $\Delta N / \Delta t$ in the number N of a population of living organisms is equal to the number of births† B minus the number of deaths D plus immigrants I minus annual emigrants E. That is,

$$\frac{\Delta N}{\Delta t} = B - D + I - E \tag{1}$$

Well now: this proposition already contains an implicit hypothesis, namely that the population of a fixed locale increases only by births or by immigration, and diminishes only by deaths or emigration. But this hypothesis will be understood by everyone without our needing to mention it specifically. Nevertheless, implied hypotheses are dangerous, and we would do well to reduce them to a minimum if not exclude them entirely.

---

* To those who see the justification of science in practical applications, it is by no means necessary to make excuses for the study of demographic analysis. Often that study allows us to obtain indirectly, by virtue of recognized relations between the attributes of a population, information for which we lack direct bases in statistical enumerations. As well, an entire industry of great importance, life insurance, bases its computations and its business upon relations closely touching on demographic analysis, if not borrowed from it.

† In the course of the present studies a very considerable number of symbols will be necessary. I have decided, not without hesitation, to preserve here the notation of my English publications, rather than to multiply still further the diversity of characters. The symbol B thus corresponds to the English word "births". Fortunately, the symbols N, t, D, I, E, in French as in English, represent by their initial letter each of the quantities in question.

Here again the question arises: would not a treatment of demographic problems that based itself on *hypotheses* in order to extract *necessary* conclusions be of doubtful practical value? We would be powerfully misled in viewing matters in that way. The conditions that present themselves in an actual population are always excessively complicated. Whoever has failed to grasp clearly the necessary relations among the characteristics of a theoretical population subject to simple hypotheses, will certainly be unable to manage in the much more complicated relations that exist in a real population. If one has wavered in the attack on a very simple problem, he will assuredly stumble in the face of very serious complications. It is for that reason that the authors who profess little respect for the application of mathematical analysis to demographic problems are also those who in their writings present us with horrible examples of the confusion that results from striving to resolve by an avalanche of words problems whose complexity imposes on us the use of the condensed language of mathematics.

On the contrary, as will be seen in what follows, mathematical analysis not only clarifies for us the fundamental relations between certain characteristics of a population, but also serves us as a guide for systematizing our inquiry. The need for such a guide is felt because of a certain peculiarity in the subject. The study of demographic analysis presents certain difficulties that one would be ill-advised to exaggerate, but about which it is well to be warned all the same. These are difficulties quite different from those, for example, that present themselves in the study of the physical sciences. In the latter it is necessary to assimilate wholly new concepts foreign to daily living, such as entropy, space-time, quanta, etc. This type of difficulty will not come to trouble us in our demographic problems. The concepts with which we will be dealing will be ideas familiar from our daily lives: the size of a population, births, ages, the number of descendants in the first generation, the second, etc.

It is thus not by their intrinsic nature that the fundamental concepts will cause us difficulties. Instead, they will occasion a very real difficulty for us by two circumstances, namely: first, their large number; and second, the complicated character of the relations that link them. Let us explain: In classical rational physics* it is possible to formulate func-

---

* These remarks no longer apply without reservation to the most recent developments in the physics of extreme dimensions. Even in classical physics the variables that enter

tional relations between variables. For example, the period T of a simple pendulum is a certain function of its length L and of the acceleration g of the force of gravity

$$T = 2\pi\sqrt{L/g} \tag{2}$$

Given, therefore, a certain length L and the gravitational constant g the period T is completely determined: it can only have a single value.*

The relations that dominate the field of demographic analysis are of a different nature. They inform us, for example, about the weight of an individual of a certain height h, not telling us that for each height h there corresponds a certain unique weight p; but at most indicating to us that, among a large number of individuals whose height lies within the limits h and $h + dh$ a certain proportion of individuals $f(h, p)\, dh\, dp$ will be found whose weight lies within the limits p and $p + dp$. Thus we no longer have here a simple functional relation, but what we will call a *probabilistic* relation; that is to say, a relation that expresses the probability that a given value of one variable will be encountered among a large number of cases with the given characteristic.

These two circumstances, the large number of variables, and the probabilistic nature of the relations that link them, lend a unique character to demographic analysis.† Among other things, they pose very inconvenient obstacles to graphic representation of the phenomena studied.

That is not to say that this method of representation will fail us entirely, and in fact we will have many occasions to make use of it. But for representing the phenomenon of demography as a whole the *usual* graphic methods are insufficient. We will see later how to get around that problem.

As for the systematic arrangement of the matter of our study, it will follow quite naturally from our analysis if we begin by considering the

---

into our formulas are actually mean values of a number of observations distributed around their means. That distribution, however, generally has a far narrower spread than in the problems that will occupy us here. In fact, the goal of the experimental method is to reduce variations to a minimum, in order to render them negligible, whereas in studies of the type that we are doing here, it is often the variations themselves that interest us.

* The square root being understood to have a positive sign.

† It is true that probabilistic relations play a role of the greatest importance in the modern theories of quanta and matter.

simplest relations, comprising a minimum of variables, and as we proceed subsequently introduce more complicated variables and relations one after the other. It is essentially this plan that we follow in our exposition.

CHAPTER 2

Relations Involving Mortality and Births

Demographic statistics is concerned primarily with human populations, and particularly with certain more or less isolated populations, as for example those of a nation or a city.

The practical problems that present themselves with respect to a population occupying a strictly limited territory, such as a city or a district, are complicated by the important role that migration across borders plays in these cases. Those complications are reduced more and more as the area included in the study expands, since emigration and immigration are plainly functions of the border periphery, whereas births and deaths are instead functions of the land area, and the ratio between the periphery and the internal area continuously decreases as the latter increases. Circumstances of politics and commerce further tend to accentuate that effect, so that for an entire country migration can in certain cases be almost negligible as a factor determining the growth of its population.

CLOSED POPULATION

By a very natural abstraction, demographic analysis envisages as a point of departure the case of a *closed* population, that is to say, a population whose numbers receive new accessions only through births and suffer losses only through deaths, immigration and emigration being excluded.

Our proposition (1) (see p. 49) then reduces simply to the formula

$$\frac{\Delta N}{\Delta t} = B - D \tag{3}$$

It will also be convenient to write

$$\frac{\Delta N}{N \Delta t} = b - d = r \tag{4}$$

the lower case characters *b*, *d*, *r* signifying the rates of birth, death and increase per capita, all being relative rates.

Among human populations for which we know the numerical values of these three quantities, the birth rate has sometimes attained a value of about 0.05 per capita, that is, 50 per thousand. In the contemporary civilized countries the rate has fallen to values of 0.017 (United States), 0.015 (England), 0.015 (France), etc.

The death rate must have been very elevated in earlier times. From the archives of the eighteenth century, in many places we encounter rates from 0.025 to 0.030; during years struck by a cholera epidemic, yellow fever, etc., the registered death rate rose to extraordinary levels, on the order of 0.045 and more. In reality, deaths must have even surpassed the registered numbers. Today, among the civilized peoples, the death rate has fallen to values around 0.010, but that is only a temporary condition. We will see later (p. 107) that we must anticipate an increase in the future, and as a permanent condition we can hardly hope to see the establishment of a death rate below about 0.014, the inverse of a hypothetical life expectancy of 70 years.

The rate of increase *r* has probably had a maximum value of about 0.03 or at most 0.04. We will see later (p. 96) that this maximum value is of particular theoretical interest.

We also remark in passing that when the rate of natural increase does not greatly exceed a value of about 0.01 per capita, we can without serious error put dN/dt in place of $\Delta N/\Delta t$ in our formulas (3) and (4), which we will do next, by writing

$$\frac{dN}{dt} = B - D \tag{5}$$

$$\frac{dN}{N\,dt} = \frac{d \ln N}{dt} = b - d = r \tag{6}$$

In fact, allowing the value 0.01 for the excess *r* of the birth rate over the death rate, at the end of one year the size of the population *N* would become

$$N' = N(1 + r) = 1.01\, N \tag{7}$$

while according to formula (6) it would be

$$N' = Ne^{r} = 1.01005\, N \tag{8}$$

essentially the same result.

Relation Between the Population Size, Annual Births, and the Life Table. Let $N(t)$ be the population and $B(t)$ the annual number of births at time t. Also let $p(a, t)$ be the probability at the moment of his birth that an individual born at time $t - a$ is still alive at time t, being then age a. This probability depends in general not only on the age a but also on the time t. However, in the present discussion we will suppose it independent of time, so that we can write more simply $p(a)$.

Clearly the number $N(t)$ of individuals living at the instant t is composed of the sum of all of those who, having been born at an instant $t - a$, have survived up to age a; this sum is obtained by assigning a all possible values from 0 to ω, the oldest possible age for the species considered.

We thus find:

$$N(t) = \int_0^{\omega} B(t - a)p(a)da \tag{9}$$

The function $p(a)$ is well known in the computations of actuaries; in their usual notation it is written l_x / l_0, the symbol l_x designating the number of survivors at age x from among a number l_0 of newborns. A table of values of l_x ordered by successive values of x is called a Life Table, and the number l_0, always arbitrarily chosen, is called the radix of the table. A power of 10, such as 100,000, is almost always employed for l_0; with that choice of radix, $p(a)$ is derived from l_x by a simple displacement of the decimal point. Stated differently, $p(a)$ corresponds to l_x in a table whose radix is unity.

Life tables have been constructed in abundance at different times since the eighteenth century* for all types of human groups, such as the principal civilized peoples, either as a whole or by geographic divisions (nations, districts, cities, etc.); an important class of life tables depicts the experience of large life insurance companies.

For species other than the human race, it is obviously much more difficult to obtain the data necessary to construct a life table, and that difficulty is especially felt where we would most want to know the facts, namely, for wild species living in their natural habitat. Still, here and there in the journals of biology some indications are found on this subject for organisms in captivity, as for example in Pearl's works on the fly Drosophila.†

In any event, for every species the values of $p(a)$ range in a continuously decreasing series from $p(0) = 1$ to $p(\omega) = 0$. As for ω, the maximum age for the individuals of the species, obviously it cannot be defined exactly.‡ However, for applications of formula (9) this does not entail any inconvenience, since the function $p(a)$ is null for every value of a greater than ω, so that we can write

$$N(t) = \int_0^\infty B(t-a)\, p(a)\, da \tag{10}$$

without being concerned about the value of ω.

Age Distribution. Formula (9) informs us about the number in the entire population, comprising individuals of all ages. By a simple revision of the limits of the integral it also indicates to us the number of individuals in any class of ages, such as individuals between the ages a_1 and a_2:

---

* For the history of the life table the reader must be referred to specialized works, for example, Dublin and Lotka, *Length of Life, A Study of the Life Table*, Ronald Press, New York, 1936.

† The idea of a life table has also been applied to certain articles in industry and commerce that must be replaced from time to time. See in this regard E. B. Kurtz, *Life Expectancy of Physical Property*, Ronald Press, New York, 1930; E. B. Kurtz and R. Winfrey, "Life characteristics of physical property," *Bulletin 103 of the Iowa Engineering Experiment Station*, Iowa State College, Ames, Iowa, 1931; A. J. Lotka, "Industrial replacement," *Skandinavisk Aktuarietidskrift*, 1933, p. 51.

‡ On this subject see the works of J. F. Steffensen and those of E. J. Gumbel.

$$N_{1,2}(t) = \int_{a_1}^{a_2} B(t-a)\, p(a)\, da \tag{11}$$

The class of individuals whose ages are comprised between ages a and $a + da$ is represented simply by the expression inside the integral, $B(t-a)\, p(a)\, da$.

It is useful to represent this number in proportion to the size $N(t)$ of the population. For that purpose, we introduce a coefficient for the age distribution $c(a, t)$, such that the number of individuals between the ages a and $a + da$ will be $N(t)\, c(a, t)\, da$. It follows that

$$N(t)\, c(a,t)\, da = B(t-a)\, p(a)\, da$$

$$c(a,t) = \frac{B(t-a)}{N(t)}\, p(a) \tag{12}$$

We remark in passing that all of the individuals in the population are between the ages of 0 and ω. Stated differently, the proportion of the population comprised within those ages is unity

$$\int_0^{\omega} c(a,t)da = 1 \tag{13}$$

Birth Rate. In formula (12) let us put

$$a = 0 \tag{14}$$

we obtain

$$c(0,t) = \frac{B(t)}{N(t)}\, p(0) \tag{15}$$

but

$$p(0) = 1 \tag{16}$$

so that

$$c(0,t) = \frac{B(t)}{N(t)} = b(t) \tag{17}$$

Thus if the function $c(a, t)$ is represented by a curve in a system of rectangular coordinates, the ordinate corresponding to the abscissa $a = 0$ represents the birth rate.

Annual Deaths. Following reasoning completely analogous to that which led us to formula (9), we obtain for annual deaths

$$D(t) = -\int_0^{\infty} B(t-a) \frac{d\,p(a)}{da} da \tag{18}$$

Death Rate. The survivors $p(a)$ at age a are reduced to $p(a) + d\,p(a)\,/\,da$ at age $a + da$. The death rate *per capita* among these $p(a)$ individuals is thus

$$\left.\begin{aligned} m(a) &= -\frac{1}{p(a)} \frac{d\,p(a)}{da} \\ &= -\frac{d \ln p(a)}{da} \end{aligned}\right\} \tag{19}$$

For the entire population, having the age distribution $c(a, t)$, the death rate will be

$$d(t) = -\int_0^{\omega} c(a,t) \frac{d \ln p(a)}{da} da \tag{20}$$

Special Case: Constant Age Distribution, Malthusian Population. If $c(a, t)$ is independent of t, we can simply write it $c(a)$. The birth rate then becomes

$$b = c(0) = \text{constant} \tag{21}$$

and

$$d = -\int_0^{\omega} c(a) \frac{d \ln p(a)}{da} da \tag{22}$$

$$= \text{constant} \tag{23}$$

$$r = b - d = \text{constant} \tag{24}$$

That is to say, the rate of increase of the population is constant

$$\frac{dN}{N\,dt} = r = \text{constant} \tag{25}$$

$$N = N_0 e^{rt} \tag{26}$$

the population grows (or declines) by the law of compound interest, the Malthusian law.

We have as well

$$B(t) = b\,N(t) \tag{27}$$

$$= b\,N_0 e^{rt} \tag{28}$$

$$= B_0 e^{rt} \tag{29}$$

annual births also obey the same law.

Next, formula (12) becomes

$$c(a) = \frac{B(t-a)}{N(t)} p(a) \tag{12}$$

$$= \frac{b\,N_0 e^{rt} e^{-ra} p(a)}{N_0 e^{rt}} \tag{30}$$

$$= b e^{-ra} p(a) \tag{31}$$

and by virtue of the relation

$$\int_0^{\omega} c(a)\,da = 1 \tag{13}$$

we have

$$1 = b\int_0^{\omega} e^{-ra}\,p(a)d(a) \tag{32}$$

$$b = 1 \Big/ \int_0^{\omega} e^{-ra}\,p(a)\,da \tag{33}$$

The coefficients b and r being known, the death rate is given immediately by the relation

$$d = b - r = \text{constant} \tag{34}$$

Annual deaths are then

$$D(t) = N(t)\,d = N_0\,e^{rt}\,d = D_0\,e^{rt} \tag{35}$$

Hence, the number in the population, annual births and annual deaths are expressed by the same Malthusian law, the law of compound interest. It is useful to have a name to designate such a population having a fixed age distribution

$$c(a) = b\,e^{-ra}\,p(a) \tag{31}$$

and progressing according to the Malthusian law, *r being some arbitrary constant*, and b satisfying the condition

$$\frac{1}{b} = \int_0^{\omega} e^{-ra}\,p(a)\,da \tag{33}$$

Such a population, computed based on an arbitrary value of r, will be called a *Malthusian* population, and its characteristics (age distribution, rates of increase, birth, death, etc.) will be Malthusian characteristics.

We will see later that a very special interest attaches to a Malthusian population having a rate of increase ρ no longer arbitrary, but defined by a certain condition of stability.*

Direct Expression for the Death Rate *d*. We have obtained expression (33) for the Malthusian birth rate; an analogous expression for the death rate results from the general formula (22) as well, which, in the present case, gives

$$d = -b \int_0^{\omega} e^{-ra}\, p'(a)\, da \tag{36}$$

$$= -\frac{\int_0^{\omega} e^{-ra}\, p'(a)\, da}{\int_0^{\omega} e^{-ra}\, p(a)\, da} \tag{37}$$

the symbol $p'(a)$ designating the derivative $dp(a)/da$.

Vital Index. The quotient b/d was introduced by R. Pearl under the name vital index. From (33) and (37) it is seen that in a Malthusian population its value is

$$\frac{b}{d} = -1 \Big/ \int_0^{\omega} e^{-ra}\, p'(a)\, da \tag{38}$$

Stationary Population. In the special case of a stationary population under the regime of a constant life table and constant age distribution, we have $r = 0$ and formulas (31), (33), (34) become

$$c(a) = b_0\, p(a) \tag{39}$$

$$b_0 = 1 \Big/ \int_0^{\omega} p(a)da = 1 / L_0 \tag{40}$$

---

* It is important to note that for a given survival function $p(a)$, we can immediately compute the Malthusian age distribution for any arbitrarily chosen r. On the other hand, to compute a *stable* age distribution, it is necessary that we know the value of ρ, whose determination presents a special problem (see pp. 114–116).

$$d_0 = b_0 = 1 / L_0 \tag{41}$$

The integral

$$L_0 = \int_0^{\omega} p(a) da \tag{42}$$

expresses the mean lifetime of N individuals (N being a sufficiently large number) who from their birth until their death find themselves constantly under the regime of mortality defined by the life table (survival function) $p(a)$. In actuarial notation we designate by $\mathring{e}_0$ this mean lifetime, also called "life expectancy at age 0," and by $\mathring{e}_x$ the mean lifetime* or life expectancy at age x. This last indicates the number of years that remain to be lived, on average, for an individual of age x; that is, in actuarial notation†

$$\mathring{e}_x = \frac{1}{l_x} \sum_{j=1}^{j=\omega} l_{x+j} \tag{43}$$

or, in the notation we have adopted

$$L_0(a_1) = \frac{\int_{a_1}^{\omega} p(a) da}{p(a_1)} \tag{44}$$

The life expectancy at age 0 is of particular interest. Following the notation in (44) we should write it $L_0(0)$. However we write it simply as

---

* It would be more exact to say the mean *survival* time at age x. However, the term *mean lifetime at age x* is established.

† Actuarial notation distinguishes the *complete* life expectancy $\mathring{e}_x$ from the curtate e_x, this latter indicating the number of *whole* years still to be lived (on average) at age x; that is to say, it does not take into account the fraction of a year that would be lived after the final anniversary of the date of birth. For us that distinction is of no interest, and we will concern ourselves only with the complete life expectancy, which we will simply call the mean lifetime, or life expectancy, at age x.

L_0, as in formula (42), since this cannot entail any misunderstanding. It will also sometimes be convenient to make use of the actuarial notation $\mathring{e}_0$ and in general $\mathring{e}_x$.

Formulas (39) and (41) show that in a stationary population under the regime of a constant life table $p(a)$, the age distribution is proportional to the quantities $l_a = l_0 p(a)$ from the life table, and that the birth and death rates, being necessarily equal to each other, are at the same time equal in value to the inverse of the life expectancy.

Let us return, however, to the more general formulas (31), (33). Those results have a particular interest, as we will have occasion to affirm at several points. It will therefore be useful to briefly recapitulate them.

Recapitulation. Under the regime of a constant life table $p(a)$, and a constant age distribution $c(a)$, the number in a population as well as the number of births increases (or decreases) following the Malthusian law, the rate of birth and the rates of death and increase are constant, and in particular

$$c(a) = b e^{-ra} p(a) \tag{31}$$

$$b = 1 \Big/ \int_0^{\omega} e^{-ra} p(a)\, da \tag{33}$$

It is important to remark that the two relations (31) *and* (33) *are not independent, and in consequence are insufficient to completely determine the two quantities b and r.* We might recall that we have expressly chosen for our starting point relations involving a minimum of variables. We will see later how to infer a complementary relationship that will serve to completely determine the values of b and r under certain conditions. But there still remain several observations to make on the case that occupies us here, and whose analytical results are expressed by formulas (31) and (33).

Statistical Application. Are formulas (31), (33), which are inferred from certain abstract hypotheses, capable of representing the state of affairs in a concrete population? In fact, three cases of such an application of these formulas have been reported. In 1907 the writer[*] made the remark that the population of England and Wales was essentially in conformity with those formulas. Four years later, in 1911, starting from the same hypotheses, L. Bortkiewicz[†] developed the same formulas and found them well suited to the population of Germany during the period from 1890 to 1900. Finally, very recently, H. Cramér applied them to the population of Sweden in 1910.[‡] Table 1 attests to the very satisfactory accord between the observed numbers and those computed from formulas (31) and (33).

Mean Age of the Population. The age distribution being

$$c(a) = b\,e^{-ra}\,p(a) \tag{31}$$

the mean age of the population of the type with which we are dealing here will be

$$A_r = \frac{b\int_0^{\omega} a\,e^{-ra}\,p(a)\,da}{b\int_0^{\omega} e^{-ra}\,p(a)\,da} \tag{45}$$

This quantity, as we will have occasion to observe below, is encountered in a number of characteristic relations in the Malthusian population. If the value of A_r must be computed for a specific value of r, the formula in the form given in (45) can be employed directly, making use of any convenient method of numerical integration that is applicable to an arbitrary function.[§] On the other hand, when it is a matter of computing a

---

[*] A. J. Lotka. "Relation between birthrates and deathrates," *Science*, 1907, *v. 26*, p. 21; ibid. "Studies on the mode of growth of material aggregates," *American Journal of Science*, 1907, *v. 24*, pp. 199, 375.

[†] *Bulletin de l'Institut International de Statistique*, v. *19* (*1*), pp. 63–138.

[‡] *Skandinavisk Aktuarietidskrift*, 1935, p. 39.

[§] That is, a function defined not by an analytical expression, but by a table of values, such as, in the present case, the function $p(a)$ defines in the life table. One finds in the manuals (for example, Whittaker and Robinson, *Calculus of Observations*) a great variety of methods for evaluating integrals of this type. The article by Bortkiewicz already cited also gives detailed indications on that computation.

Table 1. Malthusian populations. Age distributions and rates of birth, death, and natural increase, per thousand

Age group	England and Wales 1871–1880[1]					
	Males		Females		Both sexes	
	computed	observed	computed	observed	computed	observed
0–5	142	139	135	132	138	136
5–10	118	123	114	117	116	120
10–15	107	110	104	104	106	107
15–20	98	99	95	95	96	97
20–25	88	87	86	91	87	89
25–35	150	144	148	149	149	147
35–45	117	112	117	115	117	113
45–55	84	84	87	87	86	86
55–65	57	59	63	61	60	59
65–75	29	31	36	35	32	33
75–∞	11	12	13	15	12	13
Birth rate b	36.47	36.92	33.74	33.73	35.08	35.28
Death rate d	22.16	22.61	20.01	19.99	21.07	21.27
Rate of increase r	14.31	14.31	13.73	13.73	14.01	14.01

Age group	Germany 1891–1900[2] Both sexes		Sweden 1910[3] Both sexes	
	computed	observed	computed	observed
0–10	244	244	218	218
10–20	198	198	185	192
20–30	170	164	155	156
30–40	131	134	129	125
40–50	101	105	107	102
50–60	78	78	86	88
60–70	51	50	65	66
70–80	23	22	40	40
80–∞	5	4	15	13
Birth rate b	35.90	36.2		
Death rate d	22.01	22.3		
Rate of increase r	13.89	13.9		

[1] A. J. Lotka.—"Relation between birth rates and death rates," *Science*, 1907, *v. 26*, pp. 21–22; "A natural population norm," *Journal of the Washington Academy of Sciences*, 1913, *v. 3*, pp. 241, 289; "The relation between birth rate and death rate and an empirical formula for the mean length of life given by William Farr," *Quarterly Publications of the American Statistical Association*, Sept. 1918, p. 121; "Farr's relation between birth rate, death rate and mean length of life," *Quarterly Publications of the American Statistical Association*, Dec. 1921, p. 998.

[2] L. von Bortkiewicz.—"Die Sterbeziffer und der Frauenüberschuss in der stationären und in der progressiven Bevölkerung," *Bulletin de l'Institut International de Statistique*, 1911, *v. 19(1)*, p. 63.

[3] H. Cramér.—"Uber die Vorausberechnung der Bevölkerungsentwicklung in Schweden," *Skandinavisk Aktuarietidskrift*, 1935, p. 35.

whole series of values of the mean age A_r corresponding to as many values of r, formula (45) is inconvenient, and the following transformation is preferable.

We expand the function e^{-ra} in a Taylor series. Setting

$$\int_0^{\omega} a^n p(a)da = L_n \tag{46}$$

we obtain

$$A_r = \frac{L_1 - rL_2 + \frac{r^2}{2!}L_3 - \cdots}{L_0 - rL_1 + \frac{r^2}{2!}L_2 - \cdots} \tag{47}$$

$$= \lambda_1 - \lambda_2 r + \lambda_3 \frac{r^2}{2!} - \cdots \tag{48}$$

In these formulas we recognize in the quantities L the moments of the survival function $p(a)$, and in λ the (Thiele) semi-invariants [cumulants] of the same function. These last are obviously functions of L, and the reader can easily assure himself that

$$\left.\begin{aligned} L_1 &= \lambda_1 L_0 \\ L_2 &= \lambda_1 L_1 + \lambda_2 L_0 \\ L_3 &= \lambda_1 L_2 + 2\lambda_2 L_1 + \lambda_3 L_0 \\ L_4 &= \lambda_1 L_3 + 3\lambda_2 L_2 + 3\lambda_3 L_1 + \lambda_4 L_0, \text{ etc.} \end{aligned}\right\} \tag{49}$$

The law by which the successive equations are formed is easily recognized, the numerical coefficients being those of the binomial series. We reserve until later the discussion of the numerical values of the terms in L and λ.

With respect to the values of the mean age A_r, computed from formula (48), it is seen in Table 2 that they are reasonably suggestive even in the case of populations differing considerably from the Malthusian type. The table makes evident a phenomenon whose practical importance is equal to its theoretical interest: the systematic aging of this population.

Table 2. Mean age of the United States population

	1800	1850	1880	1890	1900	1910	1920	1930	1950	2000
1. Males			25.30	26.29	26.93	27.79	28.62	29.89		
2. Females			25.01	25.91	26.47	27.27	28.11	29.69		
3. Malthusian model	22.18	23.12	24.70	25.46	26.35	27.34	28.38	29.41	31.31	34.06
4. Malthusian model			24.06	24.79	25.67	26.96	28.38	30.17		

Note.— 1 and 2. Observed values. — 3. Values computed from formula (48) based on (a) the value of r corresponding to the date in question according to the growth curve of the United States (see p. 97); and (b) the life table for white females, 1919–1920. — 4. Values computed in the same way but based on a life table conforming to a date as close as possible to the date indicated in the heading. The numbers for the years 1950 and 2000 are computed based on the population and its rate of increase under the Verhulst–Pearl hypothesis.

We also remark that for

$$r = 0 \tag{50}$$

we have

$$A_0 = \lambda_1 \tag{51}$$

so that λ_1 is the mean age of a stationary population.

Variations in the Age Distribution. To understand the sense of formula (33) well, it is useful to construct a few curves based on formula (33), choosing convenient values for r and the function $p(a)$. Taking the values of $p(a)$ from the 1929–1931 American life table, and giving r the values 0.00 and 0.01 successively, we obtain the two curves of Figure 1. What strikes the eye is the relative predominance of children in the increasing population and, by contrast, the relative predominance of aged persons in the stationary population. In passing from $r = r_1$ to $r = r_2$ the

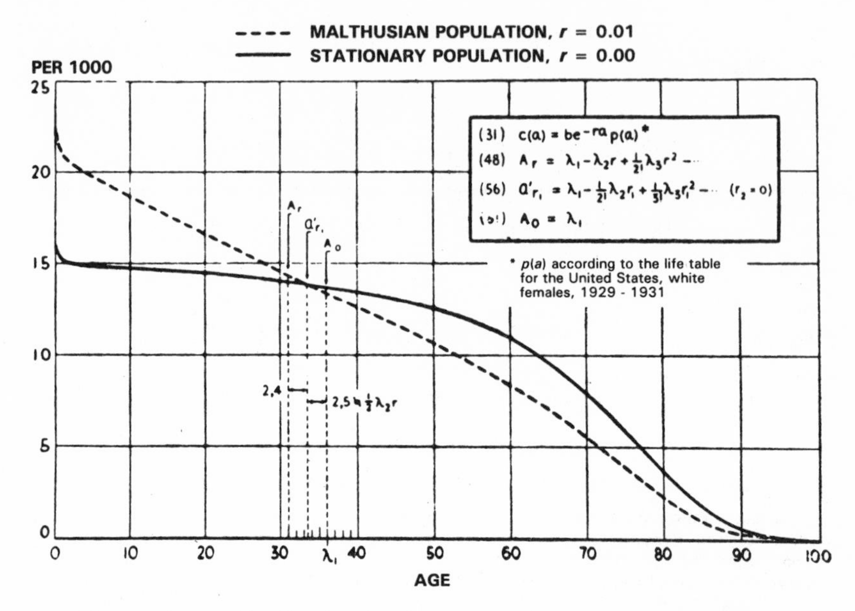

Figure 1. Age distribution [Malthusian and stationary populations].

curve *tilts* so to speak on a pivot, whose position we can find, namely the age a' satisfying the equation

$$b_1 e^{-r_1 a'} p(a') = b_2 e^{-r_2 a'} p(a') \tag{52}$$

thus

$$\frac{e^{-r_2 a'}}{e^{-r_1 a'}} = \frac{b_1}{b_2} = \frac{\int_0^{\omega} e^{-r_2 a} p(a)\,da}{\int_0^{\omega} e^{-r_1 a} p(a)\,da} \tag{53}$$

then, in view of (64), p. 71,

$$e^{-(r_2 - r_1)a'} = \frac{e^{-(\lambda_1 r_2 - \frac{1}{2}\lambda_2 r_2^2 + \cdots)}}{e^{-(\lambda_1 r_1 - \frac{1}{2}\lambda_2 r_1^2 + \cdots)}} \tag{54}$$

$$= e^{-(r_2 - r_1)\lambda_1 + \frac{1}{2}(r_2^2 - r_1^2)\lambda_2 - \cdots} \tag{55}$$

which gives the age a'

$$a' = \lambda_1 - \frac{1}{2!}(r_2 + r_1)\lambda_2 + \frac{1}{3!}(r_2^2 + r_1 r_2 + r_1^2)\lambda_3 - \cdots \tag{56}$$

corresponding to the intersection of the age distribution curves $b_1\, e^{-r_1 a}$ $p(a)$ and $b_2\, e^{-r_2 a}\, p(a)$.

Instantaneous Center of Tilt. Rather than jumping abruptly from r_1 to r_2, let us see what happens when r varies from r_1 to r_2 taking all intermediate values in succession. The momentary center of tilt then changes from moment to moment. We can find the position of the center a' corresponding to a given value of r on putting

$$\frac{\partial c(a)}{\partial r} = \frac{\partial}{\partial r}\left(be^{-ra}\, p(a)\right) = 0 \tag{57}$$

or directly using formula (56). The instantaneous center of tilt is clearly the limiting value of a' when r_2 approaches infinitesimally near to r_1. This limiting value is thus obtained very simply by putting $r_2 = r_1$ in formula (56), which gives

$$a' = \lambda_1 - \lambda_2\, r + \frac{1}{2}\lambda_3\, r^2 - \cdots \tag{58}$$

$$= A_r$$

Consequently, the instantaneous center of tilt of the curve $c(a)$ has as its abscissa the mean age A_r of the Malthusian population growing at the rate of increase r. As r passes from one value r_1 to another r_2, this instantaneous center advances along a curve that corresponds to the location of A_r.

These results have a certain practical interest. The entire trajectory of A_r between its extreme values corresponding to $r_1 = 0.01$ and $r_2 = 0$ extends only from 30 years to 40 years. The center of tilt thus rests well within the limits of the productive period of life. Every decrease in the value of r thus brings with it, for a Malthusian population, a decrease in the classes at younger ages and an augmentation in the classes at older ages. One result of this is that the part of a Malthusian population comprised within the limits of the productive period is relatively stable, as is seen in Table 3.

Table 3. Malthusian population. Age distribution (percent) computed for four values of r and four life tables

Rate of increase per capita r	England and Wales*	United States**		
	1838–1854	1901	1919–1920	1929–1931
		Ages 0–20		
− 0.010	27.40	24.82	23.68	22.28
0.000	35.10	32.33	31.11	29.58
+ 0.010	43.22	40.54	39.11	37.52
+ 0.019			46.48	
		Ages 20–50		
− 0.010	41.35	40.78	39.95	39.68
0.000	41.43	41.52	41.05	41.07
+ 0.010	39.92	40.73	40.45	40.76
+ 0.019			38.60	
		Ages 20–65		
− 0.010	59.56	60.04	59.76	60.18
0.000	55.98	57.18	57.32	58.04
+ 0.010	51.04	52.91	53.19	54.17
+ 0.019			48.40	
		Ages 65–∞		
− 0.010	13.10	15.18	16.62	17.62
0.000	8.92	10.49	11.57	12.38
+ 0.010	5.81	6.93	7.70	8.31
+ 0.019			5.12	

* Female sex
** Female sex, white race

The relative stability of the productive age class is also shared by non-Malthusian populations, as the observations show.

Relation between the Birth Rate b and Rate of Increase r. Formula (33) which expresses the relation between b and r is inconvenient for computations. If it is a matter only of a single case, given some value of r, the value of b is computed by evaluating the integral using one or another of the well known methods for integrating arbitrary functions. If, on the other hand, we desire to trace the form of the relation between b and r by computing a whole series of values of b corresponding to a series of values of r, the form of the relation (33) will render those computations very laborious. A more practical method results from the relation that we already know

$$A_r = \frac{\int_0^{\omega} a\, e^{-ra}\, p(a)\, da}{\int_0^{\omega} e^{-ra}\, p(a)\, da} \tag{45}$$

Giving the expression the form

$$A_r = -\frac{1}{1/b}\frac{\partial}{\partial r}1/b \tag{60}$$

$$= \frac{\partial}{\partial r}\ln b \tag{61}$$

we obtain

$$\ln\frac{b}{b_0} = \int_0^r A_r\, dr \tag{62}$$

$$= \lambda_1 r - \frac{1}{2!}\lambda_2 r^2 + \cdots \tag{63}$$

and finally

$$b = b_0\, e^{\lambda_1 r - \frac{1}{2}\lambda_2 r^2 + \cdots} \tag{64}$$

Moments L of the Survival Curve: Numerical Values. The integrals L_n defined by (46) are called the moments of order n of the function $p(a)$, that is, of the survival curve.

The moment of order zero

$$L_0 = \int_0^{\omega} p(a)\,da = \mathring{e}_0 \tag{42}$$

is as we have already noted, the *life expectancy*, which is given in most life tables. We can therefore find the value of L_0 directly, without any computation.

Table 4. Life expectancy for various countries and periods

Country, period	Life expectancy (years) males	females	Country, period	Life expectancy (years) males	females
France			Italy		
1817–1831	38.30	40.80	1876–1887	35.08	35.40
1840–1859	39.30	41.00	1881–1882	35.16	35.65
1861–1865	39.10	40.55	1891–1900	42.83	43.17
1877–1881	40.83	43.42	1899–1902	42.59	43.00
1898–1903	45.74	49.13	1901–1910	44.24	44.83
1908–1913	48.50	52.42	1910–1912	46.57	47.33
1920–1923	52.19	55.87	1921–1922	49.27	50.75
			1930–1932	53.8	56.0
Germany					
1871–1880	35.58	38.45	New Zealand		
1881–1890	37.17	40.25	1931	65.04	67.88
1891–1900	40.56	43.97			
1901–1910	44.82	48.33	Sweden		
1910–1911	47.41	50.68	1755–1776	33.20	35.70
1924–1926	55.97	58.82	1816–1840	39.50	43.56
1933	59.8	62.6	1841–1845	41.94	46.60
			1846–1850	41.38	45.59
England and Wales			1851–1855	40.51	44.64
1838–1854	39.91	41.85	1856–1860	40.48	44.15
1930–1932	58.74	62.88	1861–1870	42.80	46.40
			1871–1880	45.30	48.60
United States			1881–1890	48.55	51.47
1901	48.23	51.08	1891–1900	50.94	53.63
1910	50.23	53.62	1901–1910	54.55	57.00
1919–1920	55.33	57.52	1911–1920	55.60	58.38
1929–1931	59.31	62.83	1921–1925	60.72	62.95
1936	60.18	64.36	1926–1930	61.19	63.33

Table 5. Values of the principal characteristics of certain life tables

LIFE TABLE	MOMENTS					
England and Wales, females	L_0	L_1	L_2	L_3	L_4	L_5
1838–1854	41.875	1.3528×10^3	6.3298×10^4	3.4827×10^6	2.1067×10^8	1.3581×10^{10}
1871–1880	44.624	1.4666×10^3	6.9005×10^4	3.8018×10^6	2.3004×10^8	1.4808×10^{10}
1881–1890	47.187	1.5680×10^3	7.3947×10^4	4.0712×10^6	2.4559×10^8	1.5761×10^{10}
1930–1932	62.884	2.2931×10^3	11.5669×10^4	6.7214×10^6	4.2425×10^8	2.8314×10^{10}
United States,[1] females						
1901	51.102	1.7406×10^3	8.4027×10^4	4.7318×10^6	2.9169×10^8	1.9113×10^{10}
1910	53.641	1.8452×10^3	8.9253×10^4	5.0198×10^6	3.0859×10^8	2.0150×10^{10}
1919–1920	57.518	2.0149×10^3	9.9179×10^4	5.6712×10^6	3.5407×10^8	2.3455×10^{10}
1929–1931	62.829	2.2607×10^3	11.2857×10^4	6.5086×10^6	4.0864×10^8	2.7177×10^{10}
	CUMULANTS					
England and Wales, females	λ_1	λ_2	λ_3	λ_4	λ_5	
1838–1854	32.305	4.6801×10^2	4.1001×10^3	-1.7558×10^5	-8.3911×10^6	
1871–1880	32.865	4.6623×10^2	3.7270×10^3	-1.7535×10^5	-8.0022×10^6	
1881–1890	33.229	4.6293×10^2	3.4382×10^3	-1.8135×10^5	-6.9177×10^6	
1930–1932	36.468	5.2016×10^2	2.7057×10^3	-2.6151×10^5	-6.2276×10^6	
United States, females						
1901	34.061	4.8415×10^2	3.6069×10^3	-2.0275×10^5	-7.7690×10^6	
1910	34.398	4.8064×10^2	3.2790×10^3	-2.0382×10^5	-6.8588×10^6	
1919–1920	35.031	4.9712×10^2	3.3647×10^3	-2.2339×10^5	-7.4241×10^6	
1929–1931	35.981	5.0159×10^2	2.8655×10^3	-2.3562×10^5	-6.2887×10^6	

[1] 1901 and 1910 data are for the 10 states comprising the original registration area; 1919–1920 are for 27 states, from Foudray; 1929–1931 are for the United States excluding Texas and South Dakota, from Dublin and Lotka, *Length of life*; Ronald Press, New York, 1936; pp. 16–17.

The current value of the life expectancy L_0 is about 60 years according to the life tables of the principal civilized peoples. In New Zealand it has attained the remarkable level of 65.04 years for men and 67.88 for women. Table 4 displays corresponding levels for several other countries. We can see therein the progress realized during the last hundred years.

As for higher order moments, their values for several life tables are indicated in Table 5. We remark that the quotient of two consecutive moments increases quite rapidly for the first terms of the series, then more and more slowly. Clearly, that quotient can never surpass the extreme age ω, which should be its limit.* We give as an example the values of the quotient L_{n+1} / L_n for several life tables (see Table 6).

Table 6. Quotients of successive moments of the survival function

$n+1$	n	L_{n+1} / L_n								
		United States		England and Wales						
		1919 1920	1929 1931	1838 1854	1871 1880	1891 1900	1901 1910	1910 1912	1920 1922	1930 1932
1	0	35.0	36.0	32.3	32.9	33.5	34.6	35.2	36.0	36.5
2	1	49.2	49.9	46.8	47.1	47.4	48.5	49.2	50.0	50.4
3	2	57.2	57.7	55.0	55.1	55.2	56.3	56.9	57.8	58.1
4	3	62.4	62.8	60.5	60.5	*	*	*	*	63.1
5	4	66.2	66.5	64.5	64.4	*	*	*	*	66.7

* These values were not computed.

Cumulants of the Survival Curve: Numerical Values. The quantities λ_n in formula (48) are called the cumulants of order n of the function $p(a)$, that is to say, of the survival curve.

The cumulant of order one

* See A. J. Lotka, "Some elementary properties of moments of frequency distributions." *Journal of the Washington Academy of Sciences*, 1931, v. *21*, p. 17.

$$\lambda_1 = \frac{L_1}{L_0} \tag{49}$$

as we have already noted, is the mean age of a stationary population. Like the moment L_0 of order zero, a value approximating the first cumulant, it can also be taken from a life table by virtue of a certain property it possesses. We will first give the presentation and the empirical verification of that property, and will then indicate its logical basis.

The property in question is expressed either by the formula

$$\lambda_1 = \mathring{e}_{\lambda_1} \tag{65}$$

or by the more exact formula

$$\lambda_1 = \mathring{e}_{\lambda_1 - 1} \tag{66}$$ *

As an example of the applicability of these two formulas, of which the second is a little more exact than the first, Table 7 gives the results of the computation for several life tables.

Accordingly, given a life table, we have the following empirical rule for approximating the value λ_1 from the mean age of the stationary population:

Casting a glance at the column of ages and that of life expectancies, we find the region where the corresponding values in the two columns are equal. Let a be the age where this concordance is observed, then

$$\lambda_1 = a$$

gives us an approximate value for λ_1.

A more exact value is obtained by searching for the region of the table where the value in the age column is equal to the value on the line immediately preceding in the column of life expectancies, and then putting

$$\lambda_1 = a'$$

---

* *Translators' note*: For the reasons Lotka cites in the discussion which follows, the two expressions are also fairly accurate for current life tables.

a' being the value determined by this means in the column of ages.

In general, by this method, which is direct and without computation, we can only establish the region of concordance within about one year. However, if we wish, we can determine this region more exactly by linear interpolation. An example taken from the 1919–1920 American life table will serve as an illustration. In the part of the table representing the female sex, white race, we find the following values,

Age	Life expectancy
34	35.27
35	34.51
36	33.76

It is seen immediately that according to the first rule, that is, using

$$\lambda_1 = a$$

the value of λ_1 must be between 34 and 35. To determine it more exactly we put

$$34 + x = 35.27 - x(35.27 - 34.51)$$

$$x = 1.27/1.76 = 0.72$$

$$\lambda_1 = 34.72$$

Following the second rule the value of λ_1 must be between 35 and 36. To determine it we put

$$35 + x = 35.27 - x(35.27 - 34.51)$$

$$x = 0.27/1.76 = 0.15$$

$$\lambda_1 = 35.15$$

Direct computation of the value of λ_1 gives

$$\lambda_1 = L_1 / L_0 = 2015/57.52 = 35.03$$

We thus see that in this case the second rule has given a result whose error is only 0.12 year. The first rule gives too low a value, but here as well the error is only 0.31 year.

Graphic Construction of λ_1. The construction that follows brings out very clearly the relation underlying the two methods cited for the

Table 7. Comparison of values of λ_1 (mean age of the stationary population) with values computed by the approximate formulas (65) and (66)

Life table		Age that is equal to its life expectancy $a = \mathring{e}_a$	Age that is equal to the life expectancy one year earlier $a = \mathring{e}_{a-1}$	Mean age of the stationary population λ_1
England and Wales				
Females	1838–1854	32.32	32.71	32.34
"	1871–1880	32.59	33.00	32.87
"	1881–1890	32.76	33.18	33.23
"	1891–1900	33.03	33.47	33.54
"	1901–1910	34.06	34.51	34.61
"	1910–1912	34.65	35.10	35.20
"	1920–1922	35.57	36.03	36.00
"	1930–1932	36.00	36.47	36.47
United States				
Males	1901	32.83	33.25	33.25
"	1910	32.76	33.19	33.35
"	1919–1920	34.06	34.49	34.48
"	1929–1931	34.11	34.22	34.25
Females	1901	33.73	34.15	34.06
"	1910	33.92	34.36	34.40
"	1919–1920	34.72	35.15	35.03
"	1929–1931	35.40	35.86	35.98
France				
Males	1920–1923	33.10	33.54	33.48
Females	1920–1923	34.87	35.31	35.12

approximate determination of λ_1. In Figure 2 the curve of $\mathring{e}_x$, the life expectancy at age x, is found according to the 1929–1931 American life table. From the origin O a line is extended at an angle of 45 degrees. That line cuts the curve of $\mathring{e}_x$ at the point P. The abscissa OQ of point P gives the value of λ_1, the mean age of the stationary population, according to the first rule.

Likewise, if in Figure 2 we extend the line O′P′ passing through a point O′ situated on the x axis 1 unit to the left of the origin, the abscissa O′Q′ of the point of intersection P′ would give the value of λ_1 according to the second rule.

Rational Basis of the Rules for Finding the Value of λ_1. The concordance of the results obtained by the two empirical rules with the results of direct computation is very satisfying. From whence does this concordance come? It is certainly not an accident. What is its logical justification?

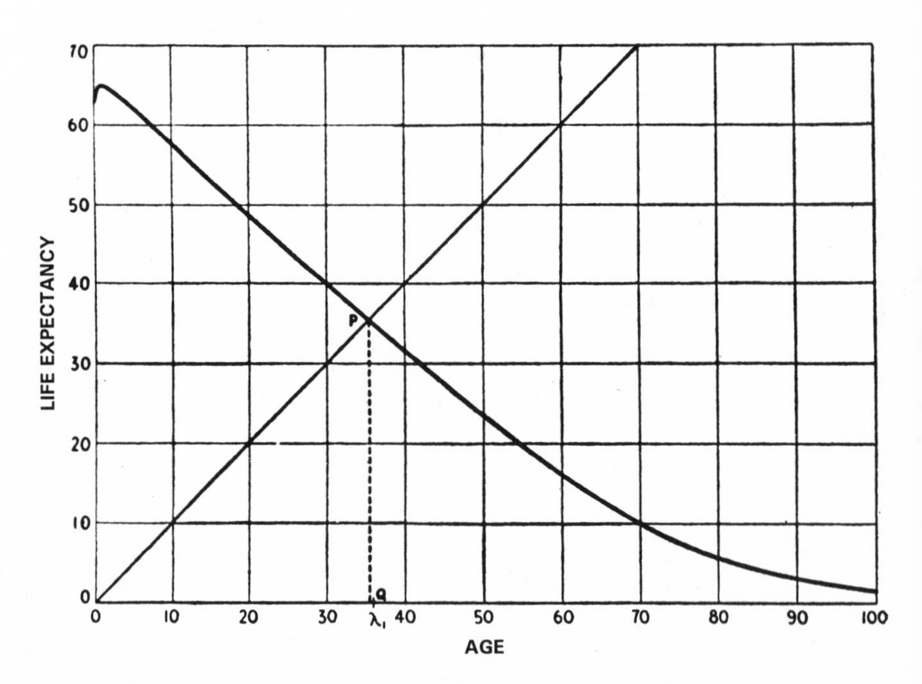

Figure 2. Graphic construction for an approximate value of λ_1, for the life table of the United States, white race, female sex, 1929–1931.

We will find the explanation by initially considering the case as presented in the simplest hypothesis about the survival function $p(a)$, "De Moivre's hypothesis," according to which this function would be linear

$$p(a) = 1 - ka \tag{67}$$

We obtain the extreme lifetime ω on putting

$$p(\omega) = 1 - k\omega = 0 \tag{68}$$

$$\omega = 1/k \tag{69}$$

thus the life expectancy will be

$$L_0 = \overset{\circ}{e}_0 = \int_0^{1/k} (1 - ka)\,da \tag{70}$$

$$= (1/2)\frac{1}{k} \tag{71}$$

On the other hand, the mean age of the stationary population will be

$$A_0 = \lambda_1 = \frac{\int_0^{1/k} a(1-ka)\,da}{\int_0^{1/k} (1-ka)\,da} = (1/3)\frac{1}{k} = (2/3)\,L_0 = (2/3)\,\overset{\circ}{e}_0 \tag{72}$$

To be sure, this result accords very poorly with the characteristics of the 1919–1920 American table, as the value

$$\lambda_1 = (2/3) \times 57.52$$

$$= 38.35$$

is very different from the true value of $\lambda_1 = 35.03$. But it is not this concordance that we are seeking. We must find for De Moivre's hypothesis the value of $\overset{\circ}{e}_{\lambda_1}$, which is equal to

$$\overset{\circ}{e}_{\lambda_1} = \overset{\circ}{e}_{1/(3k)} = \frac{\int_{1/(3k)}^{1/k} p(a)\,da}{p(1/(3k))} = \frac{\int_{1/(3k)}^{1/k} (1-ka)\,da}{1-k(1/(3k))} \tag{73}$$

$$= \frac{1}{3k} = \lambda_1 \tag{74}$$

Thus, for De Moivre's hypothesis, it is formula (65) which is exact. It will be recognized that it also gives a reasonably satisfactory approximation in the case of real life tables, as topologically the geometric properties of the survival function $p(a)$ do not differ too greatly from De Moivre's curve.

Figure 3 represents De Moivre's survival curve in rectangular coordinates, the abscissas giving the age and the ordinates the value of $p(a)$. The extreme lifetime ω forms the base of the triangle ABC, whose height AC, corresponding to $p(0)$, is equal to 1 on the scale chosen for the ordinates.

The mean age of the stationary population is represented by the abscissa of the center of gravity of the triangle, since that age is defined by

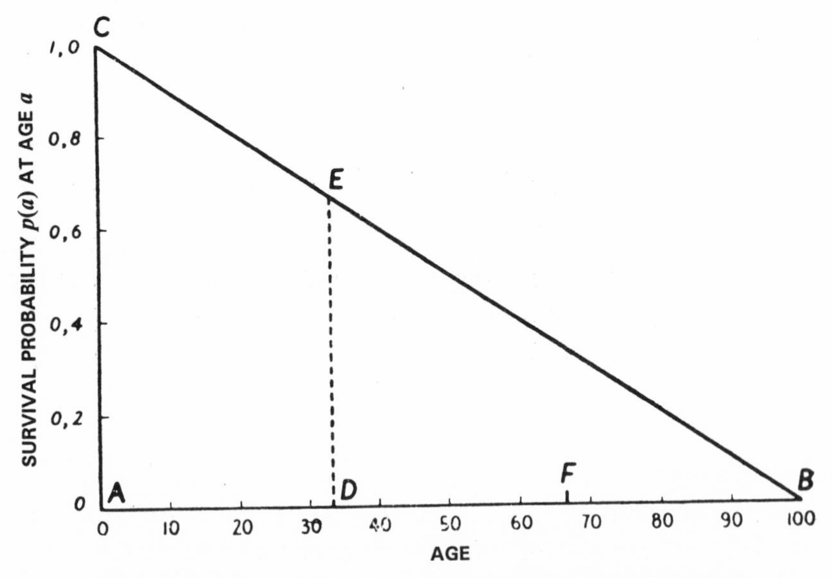

Figure 3. Survival curve according to De Moivre's hypothesis.

the formula

$$A_0 = \text{AD} = \frac{\int_0^{\omega} a\, p(a)\, da}{\int_0^{\omega} p(a)\, da} \tag{75}$$

Now, the distance from the center of gravity of a triangle to its base is equal to $1/3$ of the height. Therefore

$$\text{AD} = (1/3)\,\text{AB} = \lambda_1$$

The same diagram gives the representation of the life expectancy at age λ_1, for we have from its definition:

$$\frac{1}{p(\lambda_1)} \int_{\lambda_1}^{\omega} p(a)\, da = \frac{\text{EDB}}{\text{ED}} \tag{76}$$

$$= (1/2)\,\text{DB} \tag{77}$$

$$= (1/3)\,\text{AB} \tag{78}$$

$$= \lambda_1 \tag{79}$$

and therein is the geometric interpretation of the relation

$$\mathring{e}_{\lambda_1} = \lambda_1$$

according to De Moivre's hypothesis. Comparing Figure 3 with Figure 4, one understands why that relation also applies in an approximate manner when $p(a)$ is defined by an actual life table.

The modification introduced in formula (65) in writing $\lambda_1 - 1$ in place of λ_1 as was done in formula (66), has the character of a purely empirical correction, valid in the case of a considerable disparity between the observed curve $p(a)$ and De Moivre's hypothetical form.

We note as well a property analogous to (65). For a stationary population with an age distribution proportional to the life table, it is

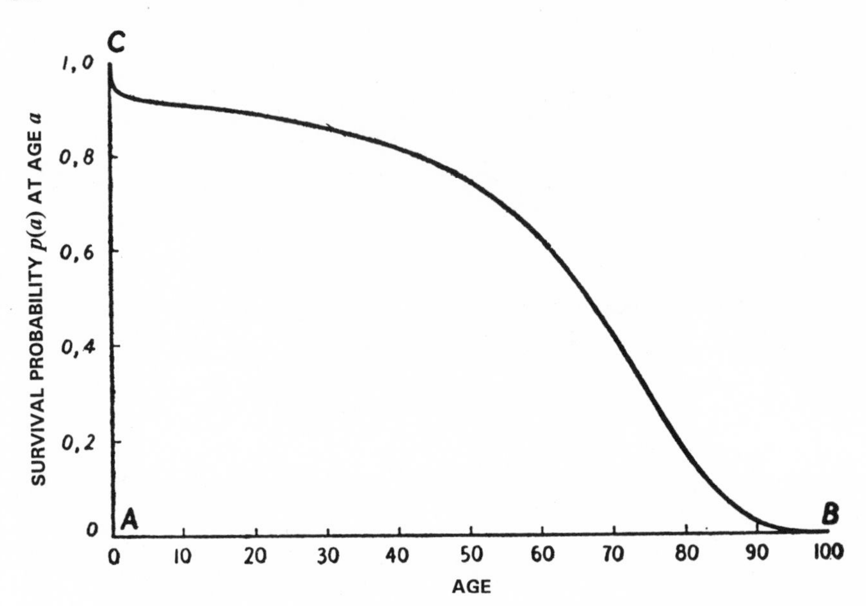

Figure 4. Actual survival curve, United States 1919–1920.

known that

$$\overline{\mathring{e}_x} = \overline{x}$$

that is, the mean of the life expectancies of all the individuals in the population is equal to their mean age.*

Higher Order Cumulants. We have seen that the values of L_0 and λ_1 can be taken from a life table without any computation, or at most by a very simple interpolation. Does this remark apply also to the cumulants λ_2, λ_3, ... etc.? In fact, here also we can manage without new computations, once the values of these constants are known for a life table that does not differ too greatly from that for which we seek the terms λ. We need to recall that the higher order terms in λ are associated with

* See for example E. F. Spurgeon. *Life Contingencies*, 1929, pp. 210–211.
Translators' note: The equality is explained by noting that the expected age of a random individual in the life table will be the mean age of the table, and the individual will have been sampled halfway through his or her lifetime.

higher powers of r and in consequence it suffices to replace the exact values λ_2, λ_3, ... from the life table in question with the corresponding values from another suitably chosen table for which the λ have been computed.

Here is a concrete illustration. Let us seek the value of b, the birth rate, according to formula (64) by applying it to the 1929–1931 United States white female life table, for which the values λ_1, λ_2, λ_3, λ_4, λ_5 have been determined directly. We find

$$b = 0.02226 \text{ for } r = +0.01$$

$$b = 0.03023 \text{ for } r = +0.02$$

What will be the result if, in place of values computed directly for 1929–1931, we employ for λ_1 the value given by the approximate rule (66) and for λ_2, λ_3, ... values taken from another life table?

We find by this means, using the values for λ_2, λ_3, ... computed according to the 1901 life table:

$$b = 0.02225 \text{ for } r = +0.01$$

$$b = 0.03050 \text{ for } r = +0.02$$

It is seen that even for the extreme value of +0.02 for r (which is hardly encountered in the statistics of modern civilized populations, with the exception of Russia), the discrepancy between the value of b computed by means of exact values of λ and that computed with their approximate values does not exceed 1 percent of the birth rate. For $r = +0.01$, always a considerable value as well, the error committed in computing b by means of approximate λ values is less than 1 / 2 per thousand.

Relation between the Birth Rate *b* and the Death Rate *d* in a Malthusian Population. Once we know the values of L_0 and the terms in λ, formula (64) enables us to compute a series of values of b corresponding to a series of values of the rate of increase r, and in consequence, a series of values of $d = b - r$. What will be the appearance of the curve so defined? Does the augmentation of the birth rate lead to the augmentation or diminution of the death rate in a Malthusian population? Graphs (Figure 1) leave us in doubt in this respect, for they tell us that in such a population there will be a relative excess of children if the birth rate is high, and a relative excess of elderly persons if the rate is low or, above all, if r is negative. However, at the two extremes of life, in early infancy as well as in old age, the death rate is very elevated. We do not know therefore, without a deeper examination of the facts, what influence the increase rate has on the character of the death rate for the entire population, comprising all ages; and in consequence we do not know, without such an examination, whether the death rate in the Malthusian population will vary in the same direction as variations in the birth rate or in the opposite direction. Computation alone is capable of informing us on this issue. That computation* gives the curves represented in Figure 5.

Minimum Value of the Death Rate. One sees that the function $d = f(b)$ initially resembles a hyperbola, the values of d diminishing with the augmentation of b and vice versa. However, for larger values of b the function $d = f(b)$ becomes increasing. Thus, for a certain value of b the function d reaches a minimum. Let us seek this value. We must put

$$0 = \frac{dd}{db} = \frac{d(r-b)}{db} = \frac{dr}{db} - 1 \tag{80}$$

hence

$$\frac{db}{dr} = 1 \tag{81}$$

Now, from (61)

$$\frac{d \ln b}{dr} = A_r \tag{82}$$

so that

---

* Constructed by means of the formula modified following the procedure on p. 71 and based on the 1930 white female life table for the United States.

$$\frac{db}{dr} = A_r b \tag{83}$$

and, finally, the condition for the minimum value of b takes the form

$$A_r b = 1 \tag{84}$$

$$A_r = 1/b \tag{85}$$

This interesting result tells us that when the death rate of a Malthusian population is at a minimum, the mean age of the population is equal to the reciprocal of the birth rate.

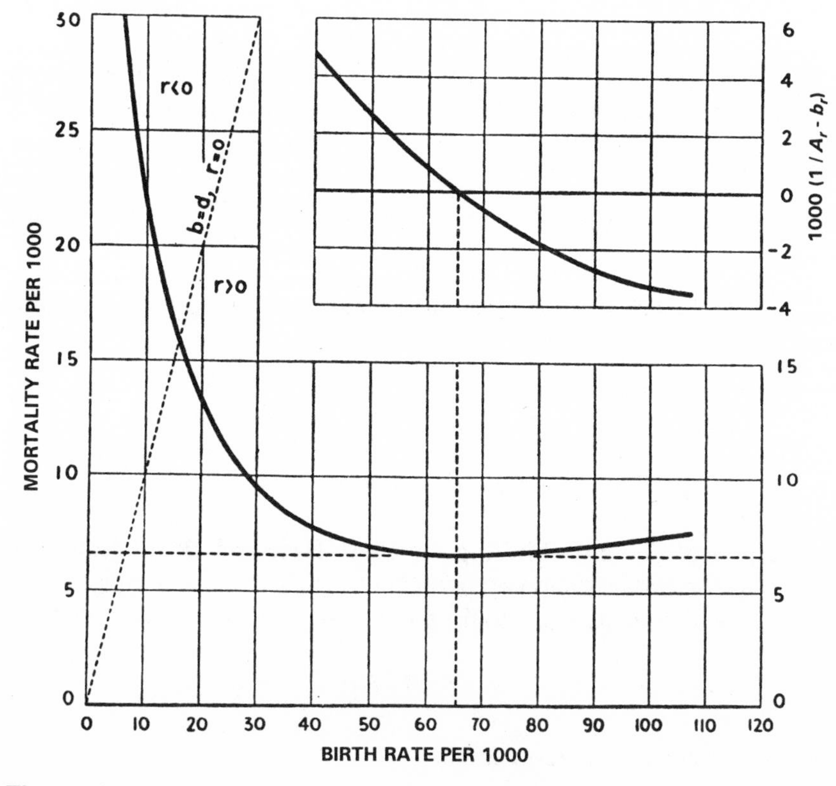

Figure 5. Malthusian population. Death rate as a function of the birth rate, with an indication of the minimum.

However, it must be said that this is a condition that will never be observed in a modern population, for the growth rate corresponding to such a state would be too high. To determine that rate we must express A_r and b in (85) by formulas (48), (41), and (64), which give

$$\lambda_1 - \lambda_2 r + \lambda_3 \frac{r^2}{2!} - \cdots = L_0 e^{-\lambda_1 r + \lambda_2 \frac{r^2}{2} - \cdots} \tag{86}$$

These formulas do not lend themselves easily to computation, however, because of the very slow convergence of the series for the values of r that interest us. It is therefore advisable to write

$$r = r' + \varepsilon \tag{87}$$

r' being an approximate solution to the problem such that ε is a very small quantity. A more exact solution is then found on putting

$$\Lambda_1 - \Lambda_2 \varepsilon + \Lambda_3 \frac{\varepsilon^2}{2!} - \cdots = L_0' e^{-\Lambda_1 \varepsilon + \Lambda_2 \frac{\varepsilon^2}{2} - \cdots} \tag{88}$$

the Λ being the cumulants and L'_0 the moment of order zero of the function

$$f(a) = e^{-r'a} p(a) \tag{89}$$

entirely analogous to the cumulants λ of $p(a)$.

Empirical Formula Linking the Birth Rate and Death Rate. We know that for $r = 0$ (stationary population) the birth rate is equal to the death rate and both are equal to the reciprocal of the life expectancy

$$d = b = 1 / L_0 \tag{41}$$

or, what amounts to the same thing,

$$(1/2)\frac{1}{b} + (1/2)\frac{1}{d} = L_0 \tag{90}$$

In general, in all practical problems b and d vary in opposite directions, and we can expect that a formula of type (90) will also be valid in an approximate manner for $b \neq d$. In fact, a formula of that type has been proposed by B. Dunlop,* evidently without knowledge of a similar formula

$$(1/3)\frac{1}{b} + (2/3)\frac{1}{d} = L_0 \tag{91}$$

cited by Newsholme and attributed by him to William Farr.†

Analytical Basis of Farr's Formula. Farr's formula clearly has a logical justification more profound than that which has been cited above. Let us first remark that a linear relation between $1/b$ and $1/d$ of type (90), (91) is equivalent to a hyperbolic relation between b and d; which recalls the comment made earlier, that for b less than $1/A_r$ the function $d = f(b)$ resembles a hyperbola. Assuming this to be the case, we ask what then is the significance of the coefficients $1/2$, $1/2$; $1/3$, $2/3$ of the empirical formulas cited? Their values have evidently been chosen arbitrarily. Is there an optimal choice, and what would its nature be?

In fact, for a Malthusian population we can establish a formula like that of Farr, in which, however, the coefficients are no longer arbitrary. Expanding the function e^{-ra} in powers of ra we obtain, from equation (33)

$$1/b = L_0 - r L_1 + \cdots \tag{92}$$

and equation (37) permits us to write

$$1/d = L_0 + r(L_0^2 - L_1) + \cdots \tag{93}$$ ‡

---

* *British Medical Journal*, 1924, p. 788.

† I have not come across the original publication.

‡ *Translators' note*: More simply, set $d = b - r$, whence using (92) we have

$$d = \frac{1}{L_0 - rL_1} - r = \frac{1 - rL_0 + r^2 L_1}{L_0 - rL_1}\,\frac{1 + rL_0}{1 + rL_0} \doteq \frac{1}{L_0 + r(L_0^2 - L_1)}$$

If the terms in r^2, etc., are negligible, upon eliminating r in the two equations (92), (93), we obtain

$$\left(1-\frac{L_1}{L_0^2}\right)\frac{1}{b}+\frac{L_1}{L_0^2}\frac{1}{d}=L_0 \tag{94}$$

In its application to the 1871–1880 female population of England and Wales (see p. 73), this formula becomes

$$0.2631\frac{1}{b}+0.7369\frac{1}{d}=44.62 \tag{95}$$

The observed value of b was 0.03373. Formula (95) gives the death rate

$$d=0.02000$$

whereas the observed value was

$$d=0.02001$$

a remarkable and undoubtedly exceptional accord.

Rounding the coefficients of $1/d$ and $1/b$ in (95) we could put

$$(1/4)\frac{1}{b}+(3/4)\frac{1}{d}=44.62 \tag{96}$$

which strongly recalls Farr's empirical formula. Once again putting

$$b=0.03373 \text{ (observed value)}$$

we obtain

$$d=0.02018$$

a very suitable accord with the observed value 0.02001.

Graph of Farr's Formula. Farr's formula admits a simple graphic representation that also applies to the more general formula (95).

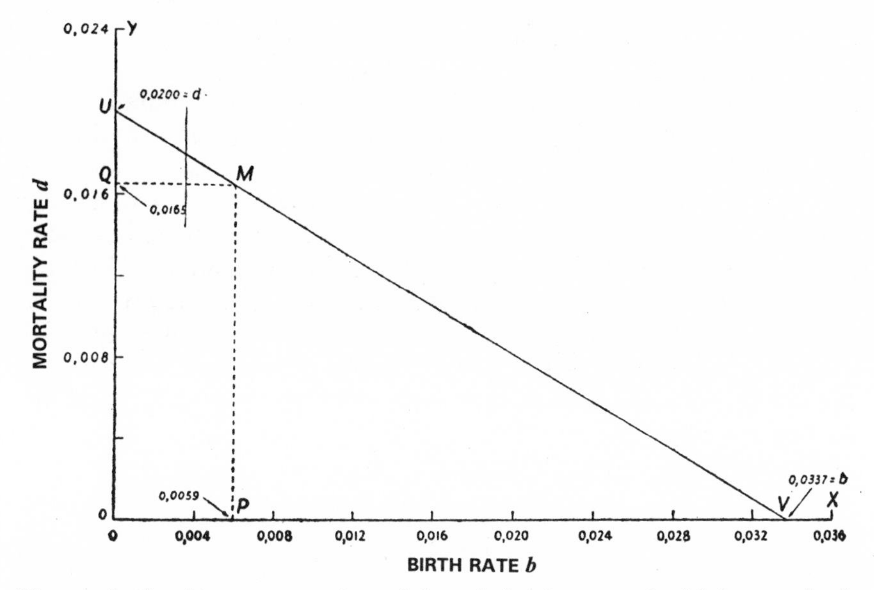

Figure 6. Graphic representation of the relation between the birth rate, death rate, and life expectancy.

Taking account of the coefficients of $1/b$ and $1/d$ in (95), we have inserted in Figure 6

$$\mathrm{OP} = \frac{1}{L_0}\left(1 - \frac{L_1}{L_0^2}\right) \tag{97}$$

$$\mathrm{PM} = \frac{1}{L_0}\left(\frac{L_1}{L_0^2}\right) \tag{98}$$

$$\mathrm{OV} = b = 0.0337 \tag{99}$$

Thus, VMU crosses the axis OY at the point U such that OU is equal to d, that is to say, in this particular case, to 0.0200.

By pivoting the line UV around the point M we obtain a succession of pairs of values b and d indicated by the successive positions of the points U and V on the scales of the vertical and horizontal axes.

POPULATION WITH VARIABLE RATES

General Considerations. Relation between the Population $N(t)$ and Annual Births $B(t)$. Thus far we have been occupied with the special case of a population whose age distribution and rates of birth, death and natural increase are constant. Even with these arbitrary and in a way artificial conditions, we have seen that our analytical model translates the structure of certain concrete populations fairly faithfully. However, such cases are rather exceptional and it is natural to seek to broaden our premises so as to approach better and in a more general manner the conditions that present themselves in concrete populations, with variable age distributions and with rates of birth, death and natural increase that are functions of time.

Toward that end we return to the fundamental equation

$$N(t) = \int_0^\infty B(t-a)\, p(a)\, da \qquad (9)$$

supposing $N(t)$ and $B(t)$ to be functions of time of some sort, susceptible, however, of being represented by the series

$$N(t) = k_0\, \varphi(t) + k_1\, \varphi'(t) + \frac{1}{2!} k_2\, \varphi''(t) + \cdots \qquad (100)$$

$$B(t) = c_0\, \varphi(t) + c_1\, \varphi'(t) + \frac{1}{2!} c_2\, \varphi''(t) + \cdots \qquad (101)$$

φ being a known function and φ', φ'', ... its successive derivatives with respect to time t. Introducing (101) in (9) we obtain

$$\begin{aligned} N(t) = {} & c_0\, \{L_0\, \varphi(t) - L_1\, \varphi'(t) + \frac{1}{2!} L_2\, \varphi''(t) - \cdots\} + \\ & c_1\, \{ \qquad L_0\, \varphi'(t) - \quad L_1\, \varphi''(t) + \cdots\} + \\ & \frac{1}{2!} c_2\, \{ \qquad\qquad L_0\, \varphi''(t) - \cdots\} + \text{etc.} \end{aligned} \qquad (102)$$

thus

$N(t) =$

$$c_0 L_0 \varphi(t) - (c_0 L_1 - c_1 L_0)\varphi'(t) + \frac{1}{2!}(c_0 L_2 - 2c_1 L_1 + c_2 L_0)\varphi''(t) \tag{103}$$

$$= \quad k_0 \varphi(t) \qquad + k_1 \varphi'(t) \qquad + \frac{1}{2!} k_2 \varphi''(t) \tag{104}$$

Comparing the terms of the same order in the right hand member of (103) and (104), we find an expression for $N(t)$ if $B(t)$ is known, or vice versa.*

Special Case. If we let

$$N(t) = K_0 \varphi(t)$$

we will then have

$$\left.\begin{aligned} c_0 &= \frac{K_0}{L_0} \\ c_0 L_1 - c_1 L_0 &= 0 \\ c_1 &= c_0 \frac{L_1}{L_0} = \lambda_1 c_0 \\ c_0 L_2 - 2c_1 L_1 + c_2 L_0 &= 0 \\ c_2 &= 2c_1 \frac{L_1}{L_0} - c_0 \frac{L_2}{L_0} \\ &= \lambda_1 c_1 - \lambda_2 c_0 \end{aligned}\right\} \tag{105}$$

---

* The problem of finding $N(t)$ when $B(t)$ is given is resolved directly by integration of the fundamental equation (9). The special case where $B(t)$ has the form of a logistic function has been treated by E. Zwinggi, *Mitteilungen des Vereins Schweizerischer Versicherungsmathematiker*, August 1929, *v. 24*, p. 94. The problem of finding $B(t)$ when $N(t)$ is given requires a different mode of treatment. Its solution had been given by Lotka, independently of the article by Zwinggi, in the *Proceedings of the National Academy of Science*, October 1929, p. 794 (Communication of Sept. 12, 1929). See also A. J. Lotka, *Human Biology*, 1931, *v. 3*, p. 459. The new method that I present here is symmetrical so to speak, and is thus equally applicable to both problems.

and so forth. We find in this way

$$\left.\begin{aligned}
c_0 &= \frac{K_0}{L_0} \\
c_1 &= \lambda_1 c_0 \\
c_2 &= \lambda_1 c_1 - \lambda_2 c_0 \\
c_3 &= \lambda_1 c_2 - 2\lambda_2 c_1 + \lambda_3 c_0 \\
c_4 &= \lambda_1 c_3 - 3\lambda_2 c_2 + 3\lambda_3 c_1 - \lambda_4 c_0 \\
c_5 &= \lambda_1 c_4 - 4\lambda_2 c_3 + 6\lambda_3 c_2 - 4\lambda_4 c_1 + \lambda_5 c_0
\end{aligned}\right\} \quad (106)^*$$

These formulas are reasonably transparent. We can, however, simplify them considerably upon giving equation (9) the form

$$N(t) = \int_0^\infty B[(t-\lambda_1)-(a-\lambda_1)]\,p(a)\,da \quad (9a)$$

We obtain in this way[†] formulas analogous to (106), on condition that we replace

$$\left.\begin{aligned}
t \text{ by } t' &= t + \lambda_1 \\
\lambda_1 \text{ by } \lambda_1' &= 0 \\
\lambda_n \text{ by } \lambda_n' &= \lambda_n \quad (n \neq 1)
\end{aligned}\right\} \quad (107)$$

and in consequence, eliminate all terms containing the factor λ_1 or c_1 in (106), which gives

---

[*] The reader will recognize here the inverse of the algorithm by which the terms in λ were derived from L (see p. 66, equation (49)), although with certain changes of sign.

[†] For a transformation of the series (101) by a change of origin and t see the Appendix, p. 201.

$$\left.\begin{aligned}
c_0' &= \frac{K_0}{L_0} \\
c_1' &= 0 \\
c_2' &= -\lambda_2 c_0 \\
c_3' &= \lambda_3 c_0 \\
c_4' &= -\lambda_4 c_0 + 3\lambda_2^2 c_0 \\
c_5' &= \lambda_5 c_0 - 10\lambda_2\lambda_3 c_0
\end{aligned}\right\} \tag{108}$$

The solution (101) to (9) then presents itself in its most characteristic form

$$\mathrm{B}(t) = \frac{K_0}{L_0}\left\{ \varphi(t') - \frac{\lambda_2}{2!}\varphi''(t') + \frac{\lambda_3}{3!}\varphi'''(t') - \frac{1}{4!}(\lambda_4 - 3\lambda_2^2)\,\varphi^{IV}(t') + \frac{1}{5!}(\lambda_5 - 10\lambda_2\lambda_3)\varphi^{V}(t') - \cdots \right\} \tag{109}$$

the time t' in the right hand member of equation (109) being counted from an origin $-\lambda_1$ prior to the time origin t in the left hand member.

If the terms in $\varphi''(t')$ and higher orders are sufficiently small in comparison to the term in $\varphi'(t')$, we will have an approximate representation of $B(t)$ in the simple form

$$B(t) = \frac{K_0}{L_0}\varphi(t') \tag{110}$$

that is, annual births $B(t)$ will follow, except for a constant term, the same law φ as the population $N(t)$, although with a displacement of the time origin toward the left at a distance λ_1, which depends solely on the function $p(a)$ and in no way on the function $\varphi(t)$.

Annual Deaths. By a procedure analogous to that which has given us $B(t)$ we can look for an expression for annual deaths $D(t)$, by starting from the equation

$$D(t) = -\int_0^\infty B(t-a)\, p'(a)\, da \tag{18}$$

where $p'(a)$ signifies the derivative of $p(a)$ with respect to age a. However, the same result is obtained more easily upon considering that

$$D(t) = B(t) - \frac{dN}{dt} \tag{111}$$

so that from (100), (101), and (105)

$$\left.\begin{aligned} D(t) &= c_0\, \varphi(t) + (c_1 - K_0)\, \varphi'(t) + \cdots \\ &= \frac{K_0}{L_0}\left\{ \varphi(t) + (\lambda_1 - L_0)\varphi'(t) + \cdots \right\} \end{aligned}\right\} \tag{112}$$

By a change of the time origin

$$t' = t + (\lambda_1 - L_0) \tag{113}$$

the second term of the right hand member cancels, and we have

$$D(t) = \frac{K_0}{L_0}\varphi(t') + R \tag{114}$$

where the remainder R contains only terms in $\varphi''(t')$ and higher orders. If those terms are reasonably small, we see that annual deaths as well as births follow the same law as the population, although with a displacement of the time origin to the right at a distance $(L_0 - \lambda_1)$, which depends only on the survival function $p(a)$ and in no way on the function φ.

Birth Rate. The birth rate $b(t)$ is inferred immediately from annual births $B(t)$ upon dividing by $N(t) = K_0\, \varphi(t)$

$$b(t) = \frac{B(t)}{N(t)} = \left\{ c_0 + c_1 \frac{\varphi'(t)}{\varphi(t)} + \frac{c_2}{2!}\frac{\varphi''(t)}{\varphi(t)} + \cdots \right\} \frac{1}{K_0} \tag{115}$$

$$= \frac{1}{L_0} \left\{ 1 + \lambda_1 \frac{\varphi'(t)}{\varphi(t)} + \frac{1}{2!}(\lambda_1^2 - \lambda_2) \frac{\varphi''(t)}{\varphi(t)} + \cdots \right\}$$

Death Rate. The population $N(t)$ being given as a function of time, the death rate $d(t)$ is obtained immediately from the relation

$$\left.\begin{aligned} d(t) &= b(t) - r(t) \\ &= b(t) - \frac{d\,N(t)}{N(t)\,dt} \end{aligned}\right\} \tag{116}$$

In the discussion of annual births and deaths we have obtained certain results of a general character without giving $\varphi(t)$ a specific form. To develop our formulas for the rates $b(t)$ and $d(t)$ further, it is necessary to take into account the form of the function φ. We will therefore consider a special case that has considerable practical interest.

Special Case: Logistic Population. We recall that the Malthusian law with which we were occupied earlier can be regarded as a particularly simple case of the general formulation

$$\frac{dN}{dt} = f(N) = a_1 N + a_2 N^2 + a_3 N^3 + \cdots \tag{117}$$

the series on the right being reduced to a single term $a_1 N$.

If in place of retaining only the first term of the series we conserve two

$$\frac{dN}{dt} = a_1 N + a_2 N^2 \tag{118}$$

we will arrive at the "logistic" law of Verhulst–Pearl. This law is known to be very closely verified by certain populations (among others that of the United States) over long intervals of time.

Let us write formula (118) in a more convenient fashion

$$\frac{dN}{N\,dt} = r_t = r_i \left(1 - \frac{N}{N_\infty} \right) \tag{119}$$

which makes evident the fact that in this case the rate of increase r is no longer constant, but decreases constantly from an *initial* value* r_i to 0; at the same time $N(t)$ increases indefinitely toward the limiting value† N_∞.

The value $N(t)$ is obtained by integrating equation (119)

$$\left.\begin{aligned} N(t) &= \frac{N_\infty}{1+e^{-r_i t}} \\ &= \frac{1}{2} N_\infty \left(1 + \tanh \frac{r_i t}{2}\right) \end{aligned}\right\} \tag{120}$$

From (120), the curve which represents the values of $N(t)$ as a function of time has the form of an elongated S.‡ As the alternative formula indicates, this curve is nothing else than a graphic representation of the hyperbolic tangent, a fact we can take advantage of in computations.

The curve of values of r_i given by formula (119) has the form of an inverted logistic, or, what amounts to the same thing, the form of a logistic whose ordinate axis reads from above downwards, as shown on the right of Figure 7.

Some Properties of the Logistic Function. In what follows we will have occasion to apply certain properties of the logistic function, which it will be useful to indicate in advance:

---

* This initial value is obviously also the maximum value (see p. 54). For the United States it is equal to 0.0314. It is probable that a value of about 0.03 to 0.04 corresponds to a physiological maximum, imposed by the fact that in a human population multiple births are rare, and the interval between successive deliveries can scarcely be less than 11 months. See A. J. Lotka, *Journal of the American Statistical Association*, 1927, pp. 166–167.

† Supposing, of course, that the population continues to obey the logistic law. This hypothesis is introduced solely to give a concrete interpretation to the constant N_∞ and not in the sense of a prediction. An extrapolation of that nature on a heroic scale would be contrary to elementary principles of science.

‡ In formulas (119) and (120) time t is counted from an origin corresponding to the center of symmetry of the curve of $N(t)$.

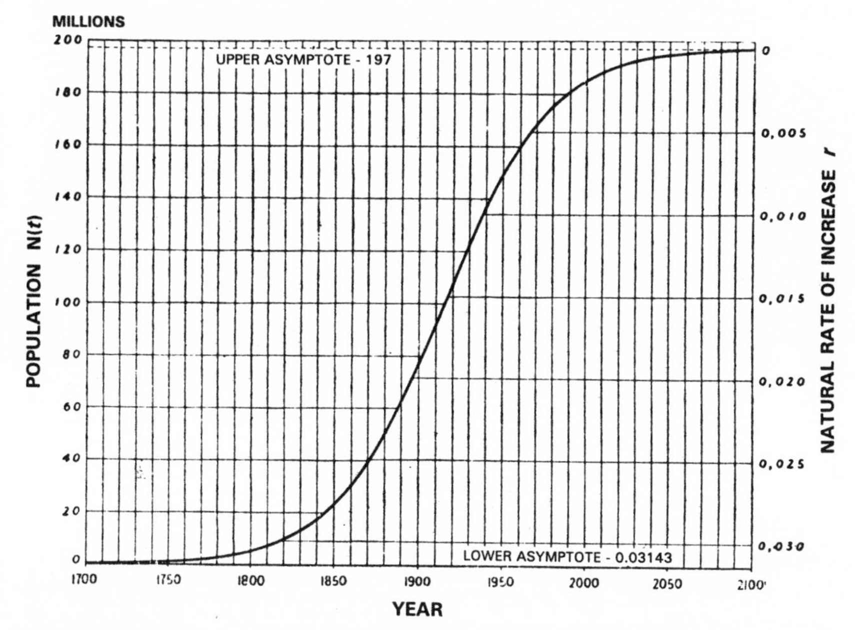

Figure 7. Logistic curve for the number N and rate of increase r of a population. Based on data for the United States. The part of the curve beyond the year 1930 is hypothetical and is not presented as a prediction.

1. Let

$$\frac{1}{1+e^{-rt}} = \psi(t) = \frac{e^{rt}}{1+e^{rt}}$$

then

$$\psi(t) + \psi(-t) = 1 \tag{121}$$

2.

$$\psi'(t) = \frac{re^{-rt}}{(1+e^{-rt})^2} = r\,\psi(t)\,\psi(-t) \tag{122}$$

so that

$$\frac{\psi'(t)}{\psi(t)} = r\,\psi(-t) \tag{123}$$

3. For the computation of terms of higher orders in formulas (103) and (104) as they apply to the case of a logistic population, it is necessary to know the successive derivatives of $\psi(t)$. They have the form

$$\left.\begin{aligned}
\psi(t) &= \frac{1}{1+e^{-rt}} \\
\psi'(t) &= \frac{re^{-rt}}{(1+e^{-rt})^2} \\
\psi''(t) &= -\frac{r^2e^{-rt}(1-e^{-rt})}{(1+e^{-rt})^3} \\
\psi'''(t) &= +\frac{r^3e^{-rt}(1-4e^{-rt}+e^{-2rt})}{(1+e^{-rt})^4} \\
\psi^{IV}(t) &= -\frac{r^4e^{-rt}(1-11e^{-rt}+11e^{-2rt}-e^{-3rt})}{(1+e^{-rt})^5} \\
\psi^{V}(t) &= +\frac{r^5e^{-rt}(1-26e^{-rt}+66e^{-2rt}-26e^{-3rt}+e^{-4rt})}{(1+e^{-rt})^6}
\end{aligned}\right\} \tag{124}$$

The law of formation for the factors in parentheses in the numerators can be expressed in a very convenient manner by an algorithm analogous to that of Pascal's triangle. In Table 8 the coefficients of the successive terms in parentheses are inscribed below the two oblique lines. The process by which the coefficient of any term is obtained is explained most easily by an example: The figure –4 in the third line is obtained by adding the two adjoining figures in the preceding line, multiplied respectively by the rank of the oblique line in which they are found. The rank is indicated by the numbers above the oblique line.* In this way we obtain the second figure in the third line (1 × –2) + (–1 × 2)

* I have borrowed this algorithm from an article (on an entirely different subject as it happens) by C. Burrau, who himself attributes it to O. Burrau. See *Skandinavisk Aktuarietidskrift*, 1934, p. 4. A table of numerical values of the first 7 derivatives of the logistic function will be found in the article by M. Merrell, *Human Biology*, 1931, *v. 3*, p. 37.

Table 8. Scheme for computing numerical coefficients in successive derivatives of the logistic function

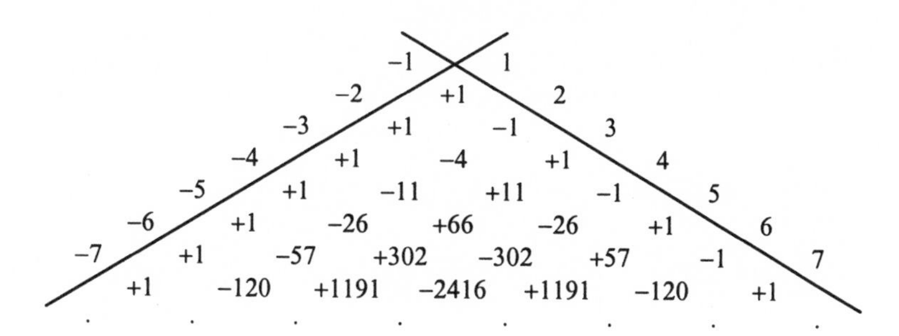

= −4. Similarly, the third figure in the last line inscribed in the table (which actually extends indefinitely) is the sum of $(-57 \times -5) + (302 \times 3) = 1191$.

Applying these properties of the logistic function, we are now ready to continue the discussion of the demographic functions.

Annual Births in a Closed Logistic Population. The expression for annual births $B(t)$ in a closed population is immediately obtained by introducing into the general formula (109) the special values corresponding to the logistic function.*

$$N(t) = \frac{N_\infty}{1 + e^{-r_i t}} = N_\infty \, \psi(t) \tag{120}$$

that is,

$$\left.\begin{aligned} \varphi(t) &= \psi(t) \\ K_0 &= N_\infty \end{aligned}\right\} \tag{125}$$

For the computations it is sometimes helpful to have the formula *in extenso* as it appears when $\varphi(t)$ is expressed by the hyperbolic tangent, for which the values are found in standard numerical tables. We then have

* It should not be forgotten that in the formula for the logistic curve written in this form, time t is counted from the "center" of the logistic.

$$B(t) = \frac{N_\infty}{L_0}\left\{\frac{1}{2}\left(1+\tanh\frac{rt'}{2}\right)+\frac{\lambda_2 r^2}{2!4}\left(\tanh\frac{rt'}{2}-\tanh^3\frac{rt'}{2}\right)-\right.$$
$$\frac{\lambda_3 r^3}{3!8}\left(1-4\tanh^2\frac{rt'}{2}+3\tanh^4\frac{rt'}{2}\right)-$$
$$\frac{(\lambda_4-3\lambda_2^2)r^4}{4!4}\left(2\tanh\frac{rt'}{2}-5\tanh^3\frac{rt'}{2}+3\tanh^5\frac{rt'}{2}\right)+$$
$$\left.\frac{(\lambda_5-10\lambda_2\lambda_3)r^5}{5!8}\left(2-17\tanh^2\frac{rt'}{2}+30\tanh^4\frac{rt'}{2}-15\tanh^6\frac{rt'}{2}\right)+\cdots\right\} \tag{126}$$

Now, the first term of the series in (109), applied to the case of a logistic population,

$$B(t) = \frac{N_\infty}{L_0}\psi(t') \tag{127}$$

itself represents a logistic curve resembling that for the number in the population, except that its center is displaced at a distance λ_1 toward the left, and in place of N_∞ its amplitude is N_∞ / L_0. As for the terms of the second order, etc., they represent essentially negligible corrections only; Figure 8, which gives the results of computations applied to numerical data for the population of the United States, shows this well. We observe in the figure that the first component, that is to say, the first term of the series, represents apart from a minor correction the whole of the function $B(t)$. In fact, to bring out the components of order 2, 3, 4, 5 well, it has been necessary to represent them in the second frame of Figure 8, on a scale twenty-five times greater than that of the first frame.

Annual Deaths in a Closed Logistic Population. Introducing into the general formula (112) the characteristics of the logistic function $N(t)$ we immediately obtain

$$D(t) = \frac{N_\infty}{L_0}\{\,\psi(t)+(\lambda_1-L_0)r_i\,\psi'(t)+\cdots\} \tag{128}$$

and, by a change in the time origin

$$\left.\begin{aligned} t' &= t + (\lambda_1 - L_0) \\ D(t) &= \frac{N_\infty}{L_0}\psi(t') + R \end{aligned}\right\} \tag{129}$$

the remainder R containing only terms in $\psi''(t)$ and higher orders.

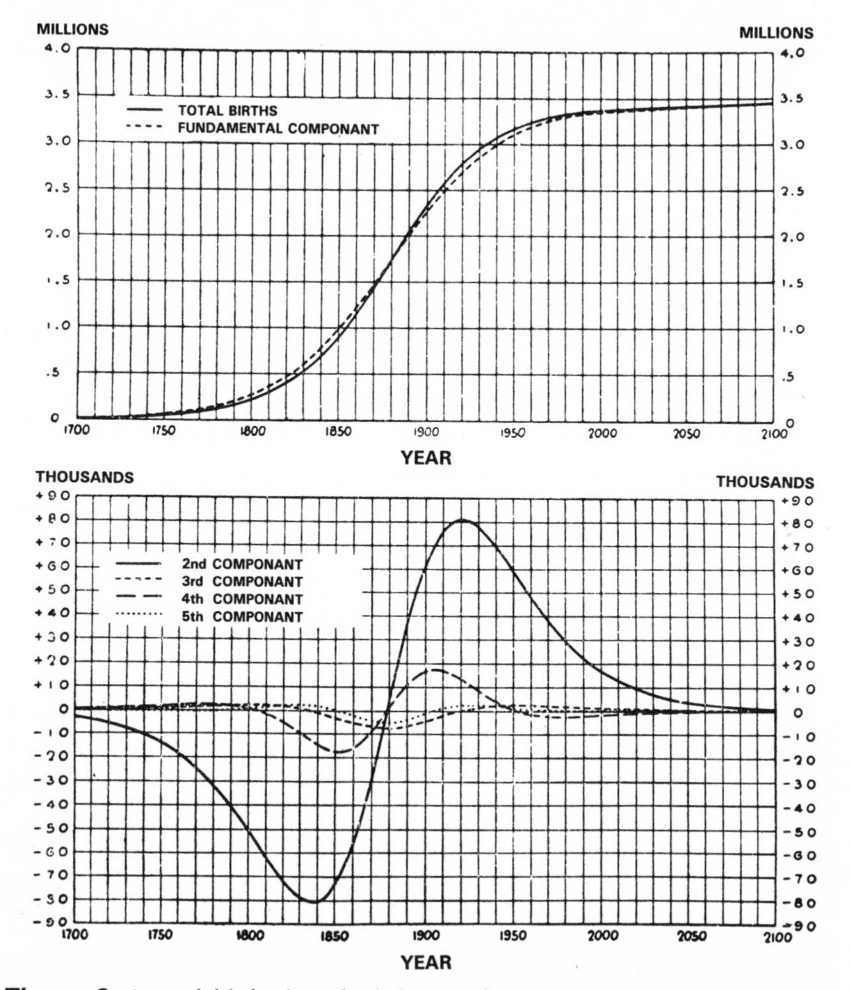

Figure 8. Annual births in a logistic population. Curve computed from the characteristics of the United States population and the 1919–1920 life table.

Birth Rate in a Closed Logistic Population. Applying to the general formula (115) the property of the logistic function

$$\frac{\psi'(t)}{\psi(t)} = r_i \, \psi(-t) \tag{123}$$

we obtain

$$b(t) = \frac{1}{L_0}\left\{ 1 + \lambda_1 \, r_i \, \psi(-t) + \cdots \right\} \tag{130}$$

In general, the two first terms of the right hand member will be insufficient by themselves to represent the function $b(t)$ even approximately, as is seen in Figure 9. However, by a change in the time origin we can give them a predominant importance. We put

$$b(t) = \frac{1}{L_0} + \frac{\lambda_1 + \tau}{L_0} r_i \, \psi(-t') + \cdots \tag{131}$$

choosing

$$t' = t + \tau$$

such that

$$\left.\begin{aligned} b_{-\infty} &= \frac{1}{L_0} + \frac{\lambda_1 + \tau}{L_0} r_i \, \psi(\infty) \\ &= b_\infty + \frac{\lambda_1 + \tau}{L_0} r_i \\ \tau &= \frac{(b_{-\infty} - b_\infty)}{r_i} L_0 - \lambda_1 \\ b(t) &= \frac{1}{L_0} + (b_{-\infty} - b_\infty)\, \psi(-t') \end{aligned}\right\} \tag{132}$$

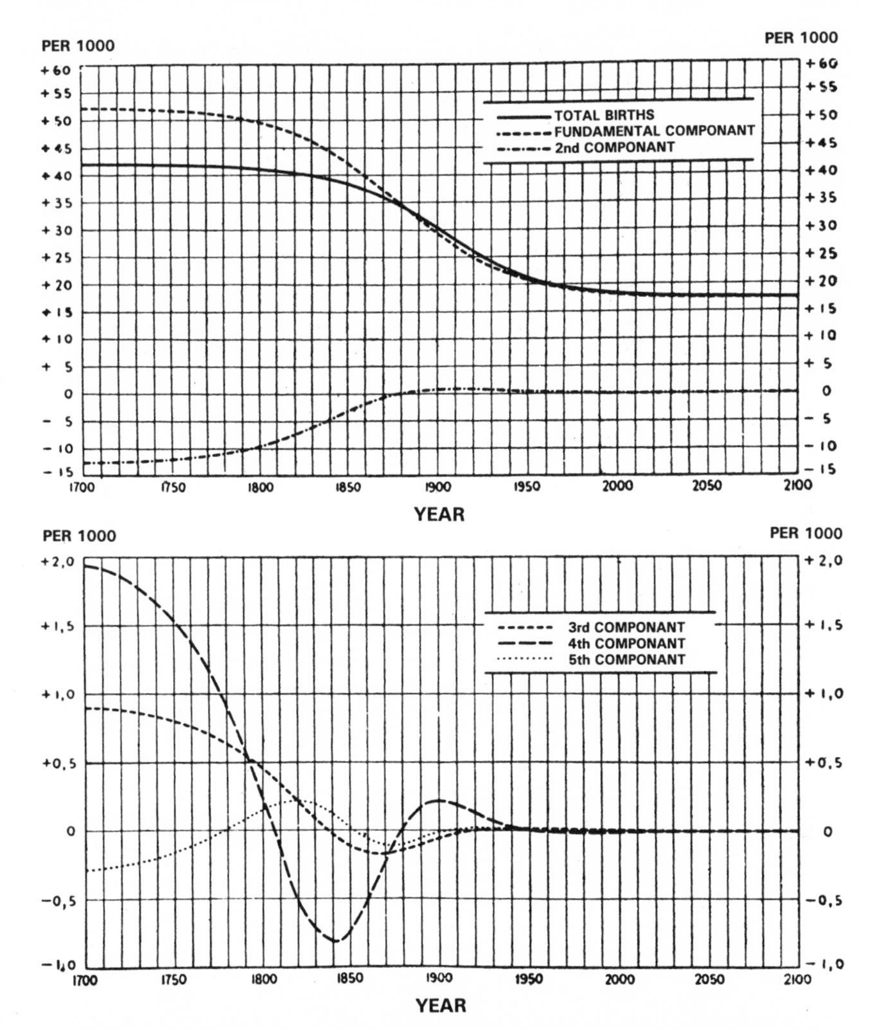

Figure 9. Birth rate in a logistic population. Curve computed from the characteristics of the United States population and the 1919–1920 life table.

With this choice for the time origin the simple formula (131) gives results almost as exact as the first six terms of formula (130), as we see in Table 9.*

---

* See also Lotka, A. J., *Annals of Mathematical Statistics*, June 1939 for a different analysis of the birth curve.

Table 9. Birth rate in a closed logistic population, based on the growth curve of the United States and the 1919–1920 life table (white females)

Year	Birth rate $b(t)$ per capita		
	Formula (130)		Formula (132)
	1st component	1st 6 terms	
1800	0.0360	0.0411	0.0410
20	0.0356	0.0404	0.0403
40	0.0348	0.0392	0.0390
60	0.0336	0.0372	0.0370
80	0.0316	0.0342	0.0341
1900	0.0290	0.0302	0.0305
20	0.0251	0.0259	0.0267
40	0.0222	0.0225	0.0234
60	0.0210	0.0202	0.0210
80	0.0195	0.0189	0.0195
2000	0.0185	0.0182	0.0185

Death Rate in a Closed Logistic Population. In an analogous manner, with a change in the time origin we find from (112)

$$\left.\begin{aligned} d(t) &= \frac{1}{L_0} + \frac{(\lambda_1 - L_0 - \tau)}{L_0} r_i \, \psi(-t') \\ \tau &= \frac{(d_{-\infty} - d_{\infty}) L_0}{r_i} + (\lambda_1 - L_0) \\ d(t) &= \frac{1}{L_0} - (d_{\infty} - d_{-\infty}) \psi(-t') \end{aligned}\right\} \tag{133}$$

Recapitulation. In a closed population $N(t)$ increasing according to the logistic law

$$N(t) = N_{\infty} \, \psi(t) = \frac{N_{\infty}}{1 + e^{-r_i t}}$$

both annual births and annual deaths increase (approximately) according to logistic laws, and the rates of birth and growth both decrease according to inverse logistic laws; however, the centers of the curves are displaced with respect to the center of $N(t)$ as indicated in the schema of Table 10 and Figures 10 and 11, whose numerical results are based on statistics for the United States.

Those relations are approximate in that we have neglected higher order terms in the series expansions. However, in the numerical example introduced by way of illustration, which is based on the logistic growth curve of the United States and 1919–1920 life table, the disparity between the approximate values of the centers of the curves and those computed by retaining 5 terms of the series reaches 7 years in only one case, that of $d(t)$. For $B(t)$ it is less than one year, and for $b(t)$ and $D(t)$ it is 2 years.

Table 10. Characteristics of a closed logistic population, computed on the basis of the growth curve of the United States and the 1919–1920 life table (white females)

Demographic function	Displacement of the center of the logistic component	Position of the displaced center	Mean value*	Position of the mean value**
$N(t)$	0.0	1914	$\frac{N_\infty}{2}$	1914
$B(t)$	λ_1	1879	$\frac{N_\infty}{2L_0}$	1879
$b(t)$	$\frac{(b_{-\infty}-b_\infty)L_0}{r_i}-\lambda_1$	1904	$\frac{b_{-\infty}+b_\infty}{2}$	1902
$D(t)$	$\lambda_1 - L_0$	1937	$\frac{N_\infty d_\infty}{2}$	1939
$d(t)$	$\frac{(d_{-\infty}-d_\infty)L_0}{r_i}+(\lambda_1 - L_0)$	1949	$\frac{d_\infty+d_{-\infty}}{2}$	1956

* That is, the mean between the two extremes corresponding to $t = \pm\infty$.
** From formulas (109), (115), conserving 5 terms.

Values of the fundamental constants.

Time origin $t = 1914$.
$N_\infty = 197$ million
$r_i = 0.0314$
$b_{-\infty} = 0.0419$
$b_\infty = d_\infty = 0.0174$
$d_{-\infty} = 0.0105$
$L_0 = 57.52$
$\lambda_1 = 35.03$

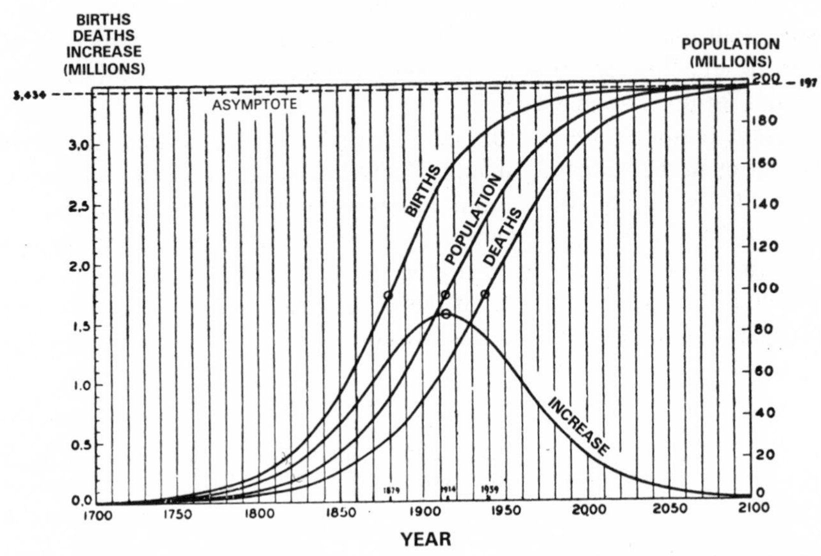

Figure 10. Curves of births, population, and annual increase in a logistic population, with an indication of the relative positions of their centers.

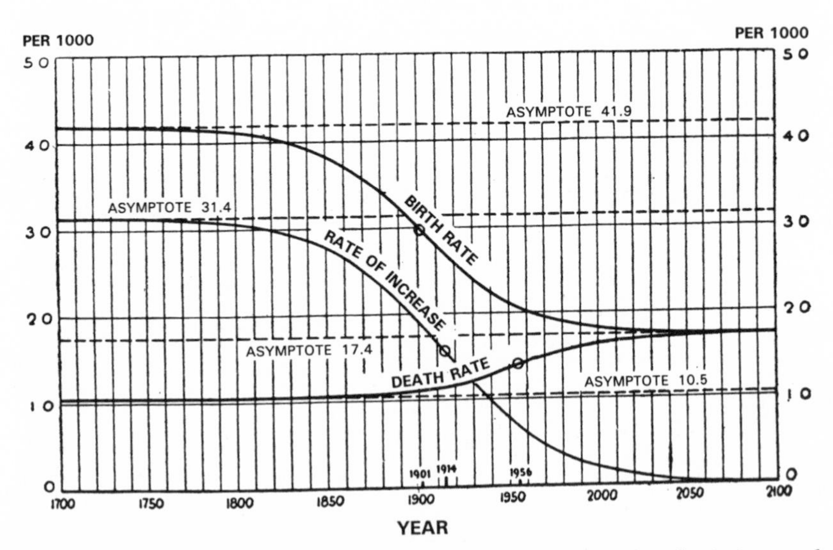

Figure 11. Curves of the rates of birth, death, and increase in a logistic population under the regime of a constant life table, indicating the relative positions of their centers.

More General Case: Logistic Population under the Regime of a Variable Life Table. An exact analysis taking account of variations in mortality presents difficulties. An approximation, on the other hand, presents itself immediately, without even changing the external form of our formulas. Provided that the function $p(a)$, which we should now write $p(a, t)$, changes only slowly with time, it is sufficient that we understand the coefficients c, as well as the cumulants λ, to be functions of time, which follows quite naturally in consequence of the secular alteration of the survival curve.

The introduction of variable mortality profoundly changes the appearance of the birth and death curves as a function of time, as is seen in Figures 11 and 12. Figure 11 represents the curves computed according to the 1919–1920 United States life table, held constant. By contrast, in the computation of Figure 12 a fluid life table has been employed, fairly closely representing the conditions corresponding to successive epochs over the period 1700 to 2100.* We immediately perceive two important differences between the two figures. The birth curve descends much more rapidly under the regime of a variable life table. What is even more striking is that the death curve under the regime of a variable life table, rather than rising continuously from an initial to a final value in the form of an elongated S, includes a period during which mortality falls fairly rapidly. This period is followed by a final stage in which the curve rises to approach the asymptotic value of the death rate. These observations are of both practical and current importance. We are today, in the United States, in the region of the minimum of the mortality curve. Up to the present we have enjoyed a continuous and quite rapid diminution of the death rate. We must anticipate an increase in that rate in the future (see p. 54).

Comparison of Theoretical Results with Experimental Data. We must emphasize that the computations and results developed in the preceding sections have only been presented by way of example, without attributing to them a strict correspondence with reality. Still, our theoretical model is constructed on bases taken from concrete demography, and we should expect that fact to be reflected in the results. Indeed, if we

---

* For details, the reader should consult the original, "The structure of a growing population," *Human Biology*, 1931, p. 481 *et. seq.*

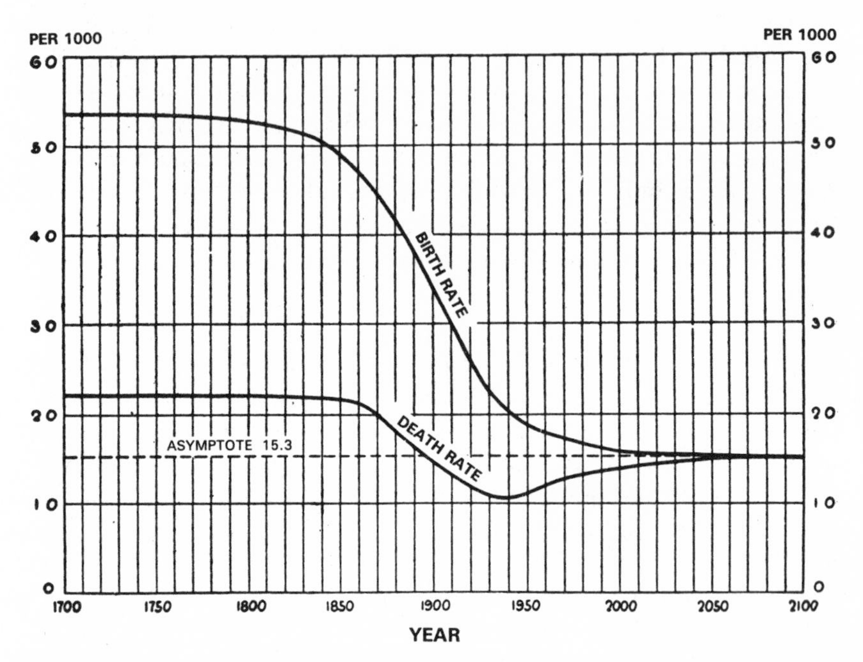

Figure 12. Curves of births and deaths in a logistic population, under the regime of a progressively ameliorating life table.

take into account the neglected factor of immigration, our theoretical results translate the concrete reality fairly faithfully. Let us see what appearance the results of our computations present from that perspective.

The Birth Rate. In the computation of the death rate according to the schema presented, we have supposed a *closed* population, that is, one increasing uniquely by way of the excess of births over deaths. In fact, the population of the United States, which has served as our example, has received very considerable numbers of immigrants over the years. In a certain sense it can be said that the quantity which we have treated as a birth rate in our computations was actually the sum of the birth and immigration rates. An approximate correction can be applied to our results by inferring from the computed values of b a quantity more or less representing the role of immigration in the growth of the population. Statistics on immigrants are quite inexact, particularly for less recent

Table 11. **Birth rate in a logistic population. Comparison of computed results with observed rates. United States. Computations based on a series of life tables (variable mortality)**

Year	Birth rate		
	Computed		Observed
	Crude	Corrected for immigration	
1915[1]	28	25	25.1
1916	28	25	25.0
1917	27	24	24.7
1918	27	24	24.6
1919	26	23	22.3
1920	26	23	23.7
1921	26	23	24.2
1922	25	22	22.3
1923	25	22	22.2
1924	24	21	22.4
1925	24	21	21.5
1926	23	20	20.7
1927	23	20	20.6
1928	22	19	19.8
1929	22	19	18.9
1930	22	19	18.9

[1] Date at which official registration of birth rates commenced in the "Birth Registration States."

periods. Basing the analysis as much as possible on the data as they exist, we arrive at the numbers presented in Table 11. We see that for the period from 1915 to 1930 there is a concordance between the computed and observed numbers that should be considered very satisfying in view of the somewhat heroic approximations that had to be made. In the years ahead, we would not contemplate even a rough concordance, the economic crisis having for the moment overturned the regular pace of events. In addition, there are reasons to believe that the growth of the American population, which as late as 1930 had very faithfully followed the logistic curve, will henceforth no longer attain the values computed according to this curve. However, it is not the objective here to make forecasts.* Our more modest task is the analysis of the facts, and more

* We remark that the integral equation (134), which we will investigate shortly (p.

particularly, the investigation of the relations between those facts.

The Age Distribution. As far as the age distribution is concerned, at least a partial correction for the effect of immigration can be made by comparing the computed numbers with only that part of the population which is of American birth. This comparison, based on a variable survival function,* is presented in Figure 13. The concordance in this case is excellent.

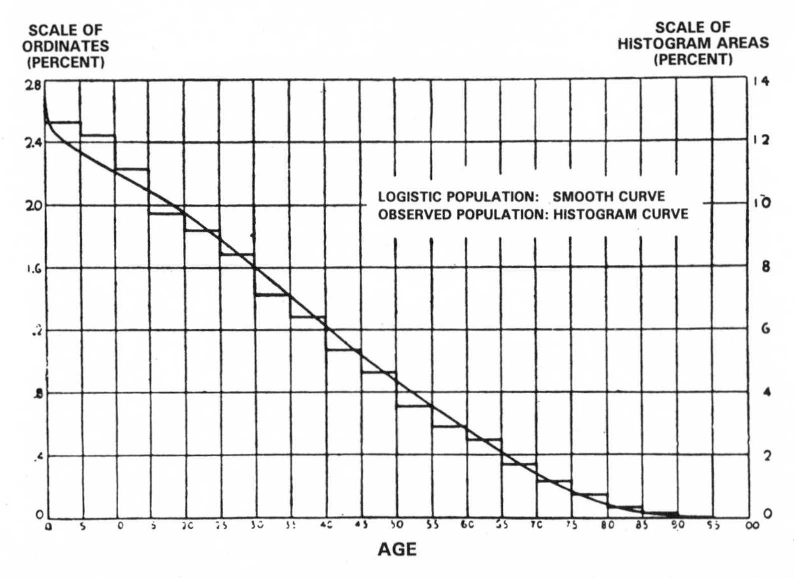

Figure 13. Comparison of the age distribution computed according to the logistic curve with the actual distribution, United States, native white females.

---

114), even permits us to compute more and more precisely the population for a series of periods in the future, as long as the values of $p(a, t)$, and a certain other function $m(a, t)$, representing fertility, are known or are given according to some hypothesis as a function of time. Computations of this type have been made by several authors with respect to various countries. For the United States they indicate a deficit below the values corresponding to the logistic for years after 1933.

* That is to say, by employing in the computations a succession of life tables corresponding to the modern progress of public health. Those computations are quite tiresome. For the details the reader is referred to the original, A. J. Lotka, "The structure of a growing population," *Human Biology*, 1931, *v. 3*, p. 459.

* * *

Following the plan indicated in the introduction, we have been occupied up to now with relations involving a minimum of variables, such as the birth rate, death rate, and age distribution.

As we have seen, there exist between these variables certain mutual relations. For example, the death rate at each age and the age distribution determine between them the overall death rate. It is evident that in an analogous manner the age distribution determines the birth rate, as the latter must depend on the count of persons comprised within the limits of the reproductive period. To trace the relations that enter into play here we pass now to the next stage of our studies, introducing a new factor into our discussion.

CHAPTER 3

Relations Involving Fertility

The formula developed in the preceding chapter

$$b = 1 \Big/ \int_0^\infty e^{-ra} p(a)\, da \qquad (33)$$

is in a certain sense incomplete. It expresses a relation between the birth rate b and the constant rate of increase r without, however, definitively fixing their values. For this a second, independent relation is required. Evidently there exists an essential factor of some sort that up to now has not entered into our computations. That factor is the fertility of the population.

Fertility and Annual Births. It will be helpful, at least for the moment, to treat our subject in its application to only one of the two sexes. For practical reasons we choose the female population.*

Annual births of daughters $B(t)$ at time t issue from mothers who at this time are of age a, comprised within the limits of the reproductive period, let us say between ages 15 and 55. The mothers themselves were born at time $(t - a)$, when annual births of girls were $B(t - a)$, so that in an interval of time dt there were $B(t - a)\, dt$ births, of which $B(t - a)\, p(a)\, dt$ survived to time t, being then between ages a and $(a + dt) = (a + da)$. Let $m(a)$ be the actual fertility† of females at age a, defined as the annual number of daughters born alive *per capita* to women of age a. The function $m(a)$, as well as $p(a)$, in the case that concerns us, is independent of time t. Then the $B(t - a)\, p(a)\, da$ women of age a will give birth

---

* The period of reproduction is more precisely defined for women than for men, and maternal parentage is almost always known, even in the case of illegitimate children.

† Potential fertility, a quantity whose measurement escapes us, will not interest us at all in this discussion.

over the course of the year to $B(t-a)\,p(a)\,m(a)\,da$ daughters, and the annual total of such births issuing from women of all ages will be

$$B(t)=\int_0^\infty B(t-a)\,p(a)\,m(a)\,da \tag{134}$$ [*]

a fundamental equation that links the annual births of daughters at time t to annual births of the mothers in earlier years.

Fundamental Equation. By its form, equation (134) recalls for us equation (9), which we encountered in the problem of a logistic population. Let us seek a solution in the form of the exponential series

$$B(t)=Q_1\,e^{r_1 t}+Q_2\,e^{r_2 t}+Q_3\,e^{r_3 t}+\cdots \tag{135}$$

Substitution gives, for $n = 1, 2, 3, \ldots$

$$1=\int_0^\infty e^{-r_n a}\,p(a)\,m(a)\,da \tag{136}$$

The coefficients r_n are determined in the same way as the roots of the equation

$$1=\int_0^\infty e^{-ra}\,p(a)\,m(a)\,da \tag{137}$$

By contrast, the coefficients Q_n depend on the initial conditions.

The Roots of the Fundamental Equation (137). General Properties. Before beginning the detailed examination of the roots of equation (137), some general remarks about their properties will be useful.

Real Root. The function $p(a)\,m(a)$ (the product of a probability and of a measure of fertility) can plainly only take positive real values. In

---

[*] We note in passing that equation (134) still remains valid if $p(a)$ and $m(a)$ are functions of time t. However, in the present discussion we treat the case in which the life table and fertility by age are fixed.

consequence, equation (137) can have but a single real root ρ, positive for

$$R_0 = \int_0^\infty p(a)\, m(a)\, da > 1$$

and negative for $R_0 < 1$:

$$\rho \gtreqless 0 \quad \text{according as} \quad R_0 \gtreqless 1 \tag{138}$$

that is to say, according to whether the number of births increases or decreases from generation to generation.

Complex Roots. Let

$$r = u + iv \tag{139}$$

be any of the complex roots of (137). Then

$$e^{rt} = e^{(u+iv)t} = e^{ut}(\cos vt + i \sin vt) \tag{140}$$

The complex roots thus introduce oscillations in the evolution of annual births. If $u < 0$, which is the ordinary case in practice, those oscillations will be damped (under the influence of the factor e^{ut}), so that the amplitude will diminish from period to period.*

In all cases we have

$$u < \rho \tag{141}$$

so that the *relative* amplitude of the oscillations with respect to the aperiodic term $Q_\rho e^{\rho t}$ continuously diminishes, and ultimately it is the aperiodic term which dominates the evolution of births, for with respect to that term all the other terms of the solution become negligible. We then have

---

* The reader will find a graphic representation of these characteristics in Figure 15, p. 131, for which a detailed discussion must, however, be postponed until later.

$$B(t) = Q_\rho \, e^{\rho t} \tag{142}$$

Important Role of the Real Root ρ: Intrinsic Rate of Increase. From the moment formula (142) is applicable, we have

$$N(t) = \int_0^\infty B(t-a)\, p(a)\, da \tag{9}$$

$$= Q_\rho \, e^{\rho t} = \int_0^\infty e^{-\rho a} p(a)\, da \tag{143}$$

$$= K e^{\rho t} \qquad (K \text{ a constant}) \tag{144}$$

We thus have, from that point onward, a Malthusian population growing or decreasing according to the constant growth rate ρ. This result is independent of the initial conditions. Provided that mortality and fertility at each age constantly maintain the given values, whatever the initial age distribution and the rates of birth and death, the growth rate ultimately approaches more and more closely to the asymptotic value ρ, the *intrinsic* value, which expresses the fundamental capacity of multiplication of the population, released from the perturbing influence of an arbitrary initial age distribution.

That is an aspect of the intrinsic rate of growth ρ that merits our particular attention. The rate interests us less as an instrument of prediction than by the fact that it expresses a real property of the population, its fundamental capacity of multiplication. The crude growth rate — the observed excess of the birth rate over the death rate — does not give us the true measure of this capacity, as both rates are influenced by the age distribution of the population, and that distribution itself depends on adventitious factors as it were, which have acted in the past. Between the crude and intrinsic rates of growth there can be an important disparity. We will return to this issue after our more detailed examination of the problem has led us to the numerical computation of the roots r.

Population with a Stable Age Distribution. This having been said, we return to the problem of the stability of the Malthusian population. We have remarked that the formulas for the birth rate and the age distribution of a Malthusian population were insufficient to completely determine its characteristics, and that an independent supplemental relation was necessary to determine the *stable* values of those characteristics. We have this independent relation in the equation

$$1 = \int_0^{\infty} e^{-ra} p(a) m(a) \, da \tag{137}$$

whose unique real root ρ completely defines the characteristics of a Malthusian population whose age distribution is stable.

These characteristics are thus

$$b_\rho = 1 \Big/ \int_0^{\infty} e^{-\rho a} p(a) \, da \tag{145}$$

$$c_\rho(a) = b_\rho \, e^{-\rho a} p(a) \tag{146}$$

$$1 = \int_0^{\infty} e^{-\rho a} p(a) m(a) \, da \tag{147}$$

The stable age distribution can actually be established in a population if the growth rate r has been practically constant over a fairly long period, as in the examples cited on p. 65. The actual distribution will diverge more or less from that stable type. If that is the case, the results contained in formulas (145), (146), (147) inform us about the final state toward which the population tends under the regime of constant fertility and mortality, without, however, indicating to us the intermediate phases through which it passes. In order to inform ourselves about the intermediate phases, we must examine the fundamental equation (137) more closely, and compute not only its real root ρ but also its complex roots $u + iv$.

The Roots of the Fundamental Equation (137): Detailed Examination. For the detailed examination and the computation of the roots of the fundamental equation (137) it is useful to introduce the moments R_n and the cumulants μ_n of the function $p(a)\, m(a) = \varphi(a)$, analogues of the moments L_n and the cumulants λ_n of the function $p(a)$.

Let

$$y = \int_0^\infty e^{-ra} \varphi(a)\, da \tag{148}$$

so that

$$\frac{dy}{dr} = -\int_0^\infty a\, e^{-ra} \varphi(a)\, da \tag{149}$$

$$= -\mathrm{A}_r \int_0^\infty e^{-ra} \varphi(a)\, da \tag{150}$$

$$= -\mathrm{A}_r\, y \tag{151}$$

$$y = R_0\, e^{-\int_0^r \mathrm{A}_r dr} \tag{152}$$

$$= R_0\, e^{-rT_r} \tag{153}$$

where we have set

$$\mathrm{A}_r = \frac{\int_0^\infty a\, e^{-ra} \varphi(a)\, da}{\int_0^\infty e^{-ra} \varphi(a)\, da} \tag{154}$$ *

$$T_r = \frac{1}{r} \int_0^r \mathrm{A}_r\, dr \tag{155}$$

We remark in passing that R_0 is the moment of order zero of the function $p(a)\, m(a)$. That parameter, by its very definition, represents the ratio between total births in two successive generations. Moreover, the quotient A_r indicates the mean age at which a mother gives birth to her daughters. We could also say that A_r is the mean difference between the

* We must distinguish A_r from A_r in formula (45), p. 64.

age of a mother and that of all of her daughters (on giving, however, to daughters born alive but afterwards dying, the age they would have been if they were still alive). Finally, T_r, being the interval during which an increase has taken place equal to the ratio between the births of two successive generations, evidently represents the mean interval between two successive generations (counted from mother to daughter, that is, along the female line of descent).

Let us return to equations (148) and, (153)

$$y = \int_0^\infty e^{-ra} \varphi(a)\, da = R_0\, e^{-r T_r}$$

$$= 1 \qquad \text{from (137)} \tag{156}$$

$$\ln R_0 - r T_r = \ln 1 \tag{157}$$

Following the model that we have used in the expansion of the mean age of a Malthusian population, we put

$$\mathrm{A}_r = \frac{R_1 - r R_2 + \frac{1}{2!} r^2 R_3 - \cdots}{R_0 - r R_1 + \frac{1}{2!} r^2 R_2 - \cdots} \tag{158}$$

$$= \mu_1 - \mu_2 r + \mu_3 \frac{r^2}{2!} - \cdots \tag{159}$$

$$T_r = \mu_1 - \mu_2 \frac{r}{2!} + \mu_3 \frac{r^2}{3!} - \cdots \tag{160}$$

In terms of the cumulants μ equation (157) becomes

$$\mu_1 r - \mu_2 \frac{r^2}{2!} + \mu_3 \frac{r^3}{3!} - \cdots - \ln R_0 = \ln 1 \tag{161}$$

Computation of the Roots of (161): Real Root. The unique real root of (161) is found upon giving to ln 1 its real value, that is, zero:*

$$\mu_1 r - \mu_2 \frac{r^2}{2!} + \mu_3 \frac{r^3}{3!} - \cdots - \ln R_0 = 0 \tag{162}$$

Approximate Values: First Approximation. In practice, r only rarely exceeds the value 0.01 or at most 0.02. We have then as a first approximation

$$r = \frac{\ln R_0}{\mu_1} \tag{163}$$

$$\doteq \frac{R_0 - 1}{\mu_1} \tag{164}$$

We can say approximately that $R_0 - 1$ is the "rate of increase *per generation*," and in the same sense r, which is obtained upon dividing that rate by the (approximate) mean interval μ_1 between two successive generations, represents the rate of increase *per year*.

An example will serve to illustrate these relations. In the white female population of the United States in 1920, we had

$$\left.\begin{array}{ll} R_0 = 1.1664 & R_0 - 1 = 0.1664 \\ \mu_1 = 28.47 & \ln R_0 = 0.1539 \end{array}\right\} \tag{165}$$

hence, from (164)

$$r = \frac{0.1664}{28.47} = 0.0058$$

---

* It is only the root in the neighborhood of $\ln R_0/\mu_1$ which counts. The other roots of equation (161) are introduced in the transformation of the fundamental equation (137) by its expansion in an exponential series, and do not correspond to the conditions of the problem.

If, to be more exact, the logarithmic formula (163) was employed, we would have found

$$r = \frac{0.1539}{28.47} = 0.0054$$

Second Approximation. Retaining the second degree term in (162), we need to solve the quadratic equation

$$\frac{1}{2}\mu_2 r^2 - \mu_1 r + \ln R_0 = 0 \tag{166}$$

which gives

$$r = \frac{\mu_1 \pm \sqrt{\mu_1^2 - 2\mu_2 \ln R_0}}{\mu_2} \tag{167}$$

In the example we have cited the computation gives

$$r = \frac{28.47 \pm \sqrt{(28.47)^2 - 2 \times 45.39 \times 0.1539}}{45.39}$$

$$= 0.0054$$

Higher Order Approximations. For the values of r encountered in practice it is only rarely necessary, *in computing the real root*, to retain terms of degrees above the second. That remark no longer applies when we investigate the complex roots, as we will see shortly. We remark in that respect that the omission of terms in the cumulants μ_3, μ_4, ..., etc., is equivalent to the admission of a Gaussian distribution for the function $p(a)\ m(a)$. Therefore, this hypothesis is permissible only for computing the real root ρ; as for the complex roots, they can be computed only by conserving the terms of orders higher than 2.

Intrinsic Birth Rate. As soon as the intrinsic growth rate ρ has been computed, we must obtain the value of the corresponding birth rate. By the application of formula (145) we obtain in our example $b = 0.02091$.

We note, however, that we can replace the substitution of $r = \rho$ in (145) by the use of an abridged and very convenient formula. The birth rate in the case of a population with a stable age distribution is obtained directly, as follows.

Combining formulas (145) and (147) we obtain

$$b = \frac{\int_0^\omega e^{-\rho a} p(a) m(a)\, da}{\int_0^\omega e^{-\rho a} p(a)\, da}$$

$$= \frac{R_0 - \rho R_1 + \frac{\rho^2}{2!} R_2 - \dots}{L_0 - \rho L_1 + \frac{\rho^2}{2!} L_2 - \dots} \tag{168}$$

Thus, as a first approximation,

$$b = \frac{R_0}{L_0} \tag{169}$$

We have as a second approximation

$$b = \frac{R_0 - \rho R_1}{L_0 - \rho L_1} = \frac{R_0}{L_0} \frac{1 - \rho \mu_1}{1 - \rho \lambda_1} \tag{170}$$

and, as successive approximations give values with alternating signs, the mean of the first and second approximations is even more advantageous, namely

$$b = \frac{R_0}{L_0} \frac{1 - \frac{1}{2} \rho (\mu_1 + \lambda_1)}{1 - \rho \lambda_1} \tag{171}$$

In our example these formulas give the following results:

Formula	(145)	(169)	(170)	(171)
b	0.02091	0.02030	0.02121	0.02075

The author will make use of this occasion to correct a strange confusion of Mr. R. Kuczynski.* That author contrasts formula (168), which he calls "the complicated formula of Lotka," with formula (145), which he attributes to Bortkiewicz. Therein is a double error. Formula (168) was introduced by Lotka only with the express aim of arriving at the approximate formula (169), which certainly is the simplest possible. As regards formula (145), it had been indicated by Lotka in 1907 for a Malthusian population with an arbitrary rate of growth r, four years before the publication of the same formula by Bortkiewicz. Its application to any population requires in general the computation of the intrinsic rate ρ, that is to say, a special method of which no trace is found in Bortkiewicz.†

From what has preceded, the value of the real root (as long as it remains within usual limits) depends almost entirely on μ_1, the mean interval between two generations, and only very slightly on the variance μ_2 of the actual fertility curve $p(a)\,m(a)$. It is practically independent of the higher cumulants of this curve.

It is entirely otherwise with the complex roots, to whose study we now pass.

Complex Roots of the Fundamental Equation. To reveal the complex roots of the fundamental equation, (137) and (161), it is necessary to write ln 1 in its complex form, that is, $\pm\, 2\pi n i$, such that

---

* R. Kuczynski, *Fertility and Reproduction*, 1932, pp. 62, 90.

† *Translators' note*: These comments reflect Lotka's irritation with Kuczynski, who did not understand the fundamental nature of Lotka's contributions to stable theory and assigned credit for key parts of it, at least in concept, to Bortkiewicz. Lotka's correction is accurate, if also strident. The various threads are disentangled in P. Samuelson, "Resolving a historical confusion in population analysis," *Human Biology*, 1976, *v. 48*, pp. 559–580; also in D. P. Smith and N. Keyfitz, editors, *Mathematical Demography: Selected Papers*, pp. 109–129. Springer–Verlag, Berlin, 1976.

$$\tfrac{1}{2}\mu_2 r^2 - \mu_1 r + \cdots + \ln R_0 = -2\pi ni \qquad (172)^*$$

$$(i = \sqrt{-1})$$

We decompose this equation into two parts, putting for the complex value of r

$$r = u + iv \qquad (173)$$

and separating the real part from the imaginary part:

$$\tfrac{1}{2}\mu_2 (u^2 - v^2) - \mu_1 u = -\ln R_0 + f(u,v) \qquad (174)$$

$$(\mu_2 u - \mu_1)v = -2\pi n + F(u,v) \qquad (175)$$

$$(n = 1, 2, 3, \ldots, \text{etc.})$$

the functions f, F being at least of the third degree in u, v and at least of the third order with respect to the cumulants μ. The factor n spans the entire range of numeric values, so that we obtain an infinite number of distinct pairs (u, v), and as many complex roots for r, whose real part u and imaginary part iv are represented by the points of intersection of the curve defined by (174), with each of the curves of the family defined by (175).

Equation (174), divided by $(1/2)\mu_2$, and completed on the right and left by the term $(\mu_1/\mu_2)^2$, takes the form

$$u^2 - 2\frac{\mu_1}{\mu_2}u + \left(\frac{\mu_1}{\mu_2}\right)^2 - v^2 = -\frac{2\ln R_0}{\mu_2} + \left(\frac{\mu_1}{\mu_2}\right)^2 + \frac{2f(u,v)}{\mu_2} \qquad (176)$$

In equations (174) and (175) let us put

---

* We could equally write $+2\pi ni$ upon making n span entirely negative values.

$$U = u - \frac{\mu_1}{\mu_2} \tag{177}$$

which gives

$$U^2 - v^2 = \left(\frac{\mu_1}{\mu_2}\right)^2 - \frac{2 \ln R_0}{\mu_2} + \chi(u,v) \tag{178}$$

$$U^2 v^2 = \frac{4\pi^2 n^2}{\mu_2^2} + \psi(u,v) \tag{179}$$

the functions χ, ψ being at least of the third degree with respect to u and v.

It will be recognized immediately that U^2 and $-v^2$ are the roots of the quadratic equation for x

$$x^2 - \left\{\left(\frac{\mu_1}{\mu_2}\right)^2 - \frac{2 \ln R_0}{\mu_2} + \chi\right\} x - \left\{\frac{4\pi^2 n^2}{\mu_2^2} + \psi\right\} = 0 \tag{180}$$

In the case of u, v, and in consequence χ, ψ sufficiently small, equation (180) can be solved by successive approximation, putting as a first approximation

$$\chi = \psi = 0 \tag{181}$$

and obtaining for u and v the provisional values u_1, v_1.

Entering these values in χ and ψ, equation (180) gives a new value of x and a new pair u_2, v_2, and so forth.

In the case of large n, the terms u, v attain considerable numerical values, which renders this method impractical. It is then necessary to operate directly on the fundamental equation (137).* However, the

---

* See A. J. Lotka, "The progeny of a population element," *American Journal of Hygiene*, 1928, *v. 8*, p. 900.

higher order roots, depending on higher order cumulants that are necessarily known in an inexact fashion, do not have much practical interest.

Special Case: The Cumulants above μ_2 Cancel. As was already stated, this condition corresponds to the function *p*(*a*) *m*(*a*) having the form of a Gaussian distribution. It is useful to examine this special case because the curves, (178) and (179), are here very easy to trace (see Figure 14). In fact, if in (178) we put

$$\chi = 0 \tag{182}$$

it will be recognized that this equation represents, in a system of rectangular coordinates, a hyperbola with the center situated at

$$\left.\begin{aligned} U &= 0 \qquad u = \frac{\mu_1}{\mu_2} \\ v &= 0 \end{aligned}\right\} \tag{183}$$

and with the axes parallel to the axes of u and v.

Similarly, if we put

$$\psi = 0 \tag{184}$$

equation (179) represents a family of hyperbolas concentric to the hyperbola (183) and having their axes inclined at 45° with respect to the axes of u and v. From a well-known property, the first hyperbola cuts each of those in the second family at 90°. The points of intersection thus correspond to the roots of the fundamental equation in the special case that we have envisioned.* It is true that this is a highly idealized case, but it is no less useful since it permits us to form an idea of the general topography of the curves representing, by their intersections, the complex roots of the fundamental equation. No doubt, if the curves were

---

* It must be noted that the curves in the upper part of Figure 14 were traced only to complete the symmetry of the design. The values of u and v corresponding to points of intersection in the two upper quadrants should be rejected, as not satisfying the condition mentioned earlier, by which there is but one real root ρ, and the real part of every complex root is less than ρ.

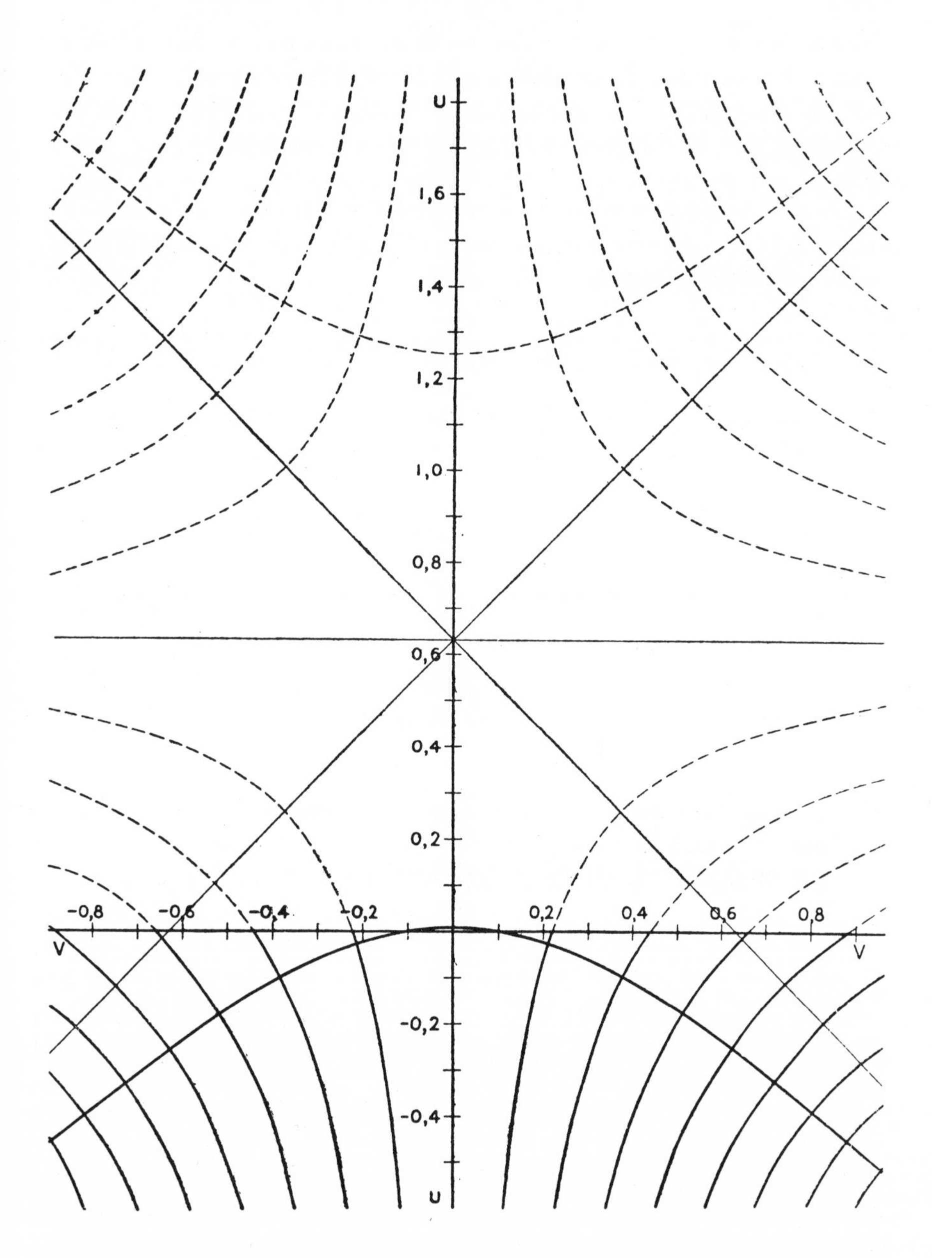

Figure 14. Topography of hyperbolic curves which, by their intersections, give the roots of the fundamental equation (137) for the case in which cumulants μ of orders higher than 2 cancel.

drawn taking into account higher order cumulants also, they would be seen to be more or less deformed with respect to the hyperbolas of Figure 14. The general topological character, however, would be conserved, at least in that the intersections would always be orthogonal.

Value Approached by the Period of the First Complex Root. Let us put $n = 1$, assuming the cumulants of higher orders to be negligible; equation (175) reduces to

$$(\mu_2 u - \mu_1)v = -2\pi \tag{185}$$

If, in addition, u is sufficiently small, one has

$$v = +\frac{2\pi}{\mu_1} \tag{186}$$

The period T of oscillation of this first component is therefore, to this degree of approximation,

$$T = \frac{2\pi}{v} = \mu_1 \tag{187}$$

that is to say, this period is equal to the mean interval between two generations from mother to daughter in the stationary population.

We will shortly see the significance of this result (see p. 131).

CHAPTER 4

The Progeny of a Population Element

The meaning of the result that was just cited will become clear after an examination of the distribution in time of the progeny of a population element.

We therefore fix our attention on a population element comprising N females, born — an essential condition — at the same moment, let us say, for example, the first of January of year zero. Let us call this element *generation zero*. When these persons attain a certain age a_1 (in practice about age 15) they commence to reproduce, giving birth to what we will call the *first generation* of daughters. This process continues up to the point that the mothers of generation zero have attained a certain age a_2 (in practice about age 55), after which all are sterile. The births of the daughters belonging to this first generation extend therefore between years a_1 and a_2, let us say 15 and 55, on the calendar that marked the year zero at the start of the observations.

At the moment that the same calendar marked the year $2a_1$ (that is, = 30) the oldest of the daughters belonging to the first generation attain age a_1 (= 15), and commence in their turn to procreate, giving birth to the *second generation* of daughters. This process in turn continues until the youngest of the daughters belonging to the first generation — those who were born in the year a_2 (= 55) — have attained the limiting age of reproduction a_2 (= 55). The calendar will then mark the date $2a_2$ (= 110). The second generation extends therefore from the year $2a_1$ to the year $2a_2$, the third from $3a_1$ to $3a_2$, and so forth, the nth from na_1 to na_2.

The distribution of births within these limits will, however, be far from being uniform. Let us look for the law of this distribution. Of N individuals of generation zero, there survive up to age a a number equal to $N\,p(a)$. These surviving women produce during the time $da = dt$ a number equal to $N\,p(a)\,m(a)\,da$ daughters (first generation). Thus, the distribution of the births $B_1(t)$ belonging to the first generation is given by the formula

$$B_1(t) = N\,p(a)\,m(a) = N\,p(t)\,m(t) \tag{188}$$

For the second generation a reasoning entirely analogous to that developed on pp. 113–114 leads directly to the relation

$$B_2(t) = \int_0^t B_1(t-a)\,p(a)\,m(a)\,da \tag{189}$$

$$B_3(t) = \int_0^t B_2(t-a)\,p(a)\,m(a)\,da \tag{190}$$

and, in general, for the $(n + 1)$th generation

$$B_{n+1}(t) = \int_0^t B_n(t-a)\,p(a)\,m(a)\,da \tag{191}$$

Analytical Property of the Distributions of Births in Successive Generations. Equation (191) permits us to immediately draw the following conclusions*

1. Whatever the integer k, the cumulant of order k of the distribution of births in the $(n + 1)$th generation exceeds that of the nth generation by μ_k, this last parameter being the cumulant of the same order in the function $p(a)\,m(a)$.
2. In consequence, the cumulant of order k of the nth generation is equal to $n\mu_k$.
3. In consequence, as n increases, the distribution of births in the nth generation approaches more and more closely to the normal (Gaussian) distribution, with the variance $n\mu_2$.†

---

* These are due to the additive property of Thiele's cumulants. See in this connection A. J. Lotka, "The progeny of a population element," *American Journal of Hygiene*, 1928, *v. 8*, p. 875.

4. The limits na_1 and na_2 of the nth generation indicate only its maximum extent. In practice almost all births will be comprised within the limits $\pm\sqrt{n\mu_2}$ around the mean year of the birth distribution in this generation.

These properties of the distribution with respect to time of births in successive generations receive a graphic interpretation in Figure 15 below, which also allows us to see why the fundamental period of oscillation T (see p. 128) is nearly equal to the mean interval between two successive generations.

Total Births at Time *t*. At any given time t births will belong in general to several contemporary generations. In fact, births will be possible for every nth generation satisfying the inequality

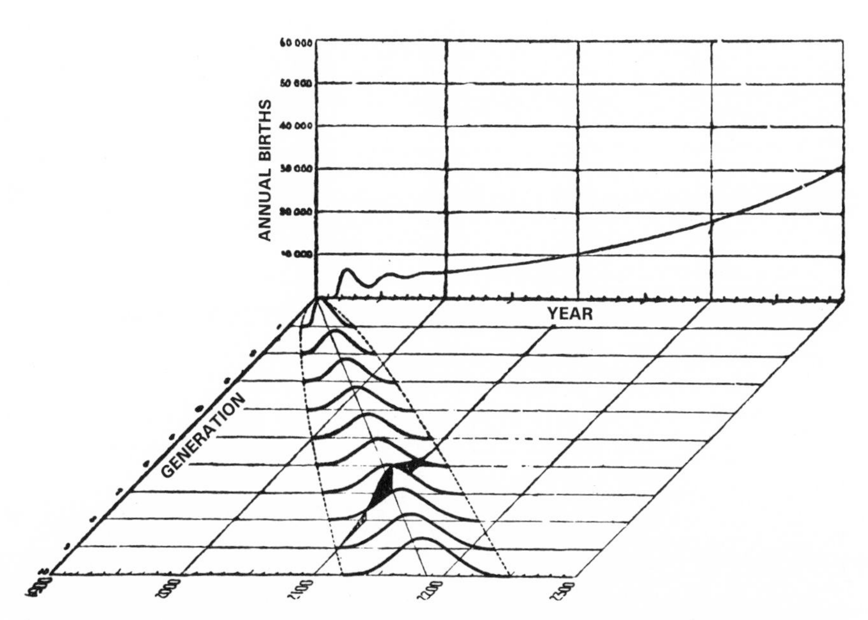

Figure 15. Distribution in time of births in a succession of generations. The curve traced on the rear face of the figure is the resultant, giving the total births for all coexisting generations.

---

† See Lotka, *loc. cit.*, and the similar paper, *Skandinavisk Aktuarietidskrift*, 1933, p. 51.

$$na_1 < t < na_2 \tag{192}$$

or what amounts to the same thing

$$\frac{t}{a_2} < n < \frac{t}{a_1} \tag{193}$$

For example, at time $t = 500$, births will be possible among all generations comprised between

$$n = \frac{t}{a_1} = \frac{500}{15} \cong 33 \tag{194}$$

and

$$n = \frac{t}{a_2} = \frac{500}{55} \cong 9 \tag{195}$$

Actually, almost all births will be comprised between much narrower limits. Let us see how the total number of annual births is composed of births belonging to diverse contemporary generations.

Let us return to equation (191) and assign n the values 1, 2, 3, ...

$$\left.\begin{aligned} B_1(t) &= B_1(t) \\ B_2(t) &= \int_0^t B_1(t-a)p(a)m(a)da \\ B_3(t) &= \int_0^t B_2(t-a)p(a)m(a)da \\ &\cdots \\ B_{n+1}(t) &= \int_0^t B_n(t-a)p(a)m(a)da \end{aligned}\right\} \tag{196}$$

The sum of these equations has the form

$$\sum_{i=1}^{i=j+1} B_i(t) = B_1(t) + \int_0^t \sum_{i=1}^{i=j} B_i(t-a)\,p(a)\,m(a)\,da \tag{197}$$

Now, if the sum Σ in the right part is extended over all the generations having members born at time t, or, stated differently, if the jth generation is the youngest that includes members born at this time,* we can suppress the last term in the sum Σ in the left part, which gives

$$\sum_{i=1}^{i=j} B_i(t) = B_1(t) + \int_0^t \sum_{i=1}^{i=j} B_i(t-a)\,p(a)\,m(a)\,da \tag{198}$$

or, very simply

$$B(t) = B_1(t) + \int_0^t B(t-a)\,p(a)\,m(a)\,da \tag{199}$$

the symbol $B(t)$ without an index designating the total number of births of all contemporary generations at time t.

We remark again that, from the moment $t = a_2$, births of the first generation have terminated, that is, $B_1(t) = 0$ and we have for $t > a_2$

$$B(t) = \int_0^\infty B(t-a)\,p(a)\,m(a)\,da \tag{134}$$

Hence we rediscover our formula (134) as a special case of the more general formula (199), namely, the case which presents itself when we consider a population of "ancient" origin, whose age exceeds a_2 years.

Distribution of Births among Different Contemporary Generations. We have remarked that at any time t all births belong to generations comprised between the limits

---

* *Translators' note*: That is, if $B_{j+1}(t)=0$, as no members of this generation are yet born. In that event $\sum_{i=1}^{i=j+1} B_i(t) = \sum_{i=1}^{i=j} B_i(t) = B(t)$, establishing expressions (198) and (199).

$$\frac{t}{a_2} < n < \frac{t}{a_1} \tag{193}$$

However, the contributions of these generations in the total mass of births will obviously be very unequal.

To determine the proportion of births belonging to any generation, let us say the $(n+j)$th, at time t, we first fix our attention on the distribution over time of the births of the $(n+j)$th generation; we allow that $(n+j)$ is a large enough number that we are permitted to suppose this distribution to be normal (Gaussian, see p. 130). We have seen elsewhere that the mean of the distribution curve corresponds to $t=(n+j)\mu_1$ and the variance is equal to $(n+j)\mu_2$, while the cumulative number of births in the entire $(n+j)$th generation is given by $N R_0^{n+j}$.

Let us designate by $\Phi(n+j,t)$ the distribution over time of births in this generation, so that $\Phi(n+j,t)\,dt$ represents the number of such births taking place at period t in an interval of time dt; we have by our hypotheses

$$\Phi(n+j,t) = \frac{R_0^{n+j}}{\sqrt{2\pi(n+j)\mu_2}} e^{-[t-(n+j)\mu_1]^2/2(n+j)\mu_2} \tag{200}$$

Put

$$t = n\mu_1 \tag{201}$$

which gives

$$\Phi(n+j) = \frac{N R_0^{n+j}}{\sqrt{2\pi(n+j)\mu_2}} e^{-j^2\mu_1^2/2(n+j)\mu_2} \tag{202}$$

For n sufficiently large and j sufficiently small, respectively, we find the simpler expression

$$\Phi(n+j) = \frac{N R_0^{n+j}}{\sqrt{2\pi n\mu_2}} e^{-j^2\mu_1^2/2n\mu_2} \tag{203}$$

On the other hand, we have seen that according to (162)

$$R_0 = e^{r(\mu_1 - \frac{1}{2}\mu_2 r + \cdots)} \tag{204}$$

thus for r sufficiently small, we have

$$R_0^j = e^{r\mu_1 j} \tag{205}$$

To simplify formula (203) let us put

$$\frac{n\mu_2}{\mu_1^2} = \delta^2 \tag{206}$$

under these conditions formula (203) becomes

$$\Phi(n+j) = \frac{N R_0^n}{\mu_1 \delta \sqrt{2\pi}} e^{-r\mu_1 j - j^2/2\delta^2} \tag{207}$$

$$= \frac{N R_0^n}{\mu_1 \delta \sqrt{2\pi}} e^{-(j - r\mu_1\delta^2)^2/2\delta^2 + r^2\mu_1^2\delta^2/2} \tag{208}$$

$$= \frac{N R_0^{n + r\mu_1\delta^2/2}}{\mu_1 \delta \sqrt{2\pi}} e^{-(j - r\mu_1\delta^2)^2/2\delta^2} \tag{209}$$

It will be realized therefore that by our hypotheses — that is, for n sufficiently large and j/n as well as r sufficiently small — the contributions of the different generations of order $(n + j)$ are distributed around that of the nth generation according to the normal (Gaussian) law, with a standard deviation

$$\delta = \frac{\sqrt{n\mu_2}}{\mu_1} \tag{210}$$

while the distribution of births with respect to time in the nth generation is itself normal with a standard deviation

$$\delta_n = \sqrt{n\mu_2} \tag{211}$$

so that

$$\delta = \frac{\delta_n}{\mu_1} \tag{212}$$

The center of the distribution of births in the $(n + j)$th generation, however, is not situated exactly at $j = 0$ but at $j = r\mu_1\delta^2 = rn\mu_2/\mu_1$.

These results lend themselves to an interesting graphic representation, given in Figure 15. In the figure, the distribution of births in successive generations is represented by the frequency curves arranged in their natural order n = 1, 2, 3, ... at a distance μ_1 one from another. In this three dimensional model the frequency curve with respect to time of any nth generation (n sufficiently large), displaced by a rotation of 90° around a vertical axis passing through its mean, becomes very nearly the frequency curve of the contributions of successive generations at period $t = n\mu_1$.

Numerical Example. The preceding formulas have been applied by way of example to a generation zero composed of 100,000 births and possessing the same characteristics as the white female population of the United States for the year 1920. Giving the parameters the following numerical values

$$\left.\begin{array}{lll} r = 0.005 & \mu_1 = 28.47 & \mu_2 = 45.39 \\ R_0 = 1.1664 & N = 100{,}000 & \end{array}\right\} \tag{213}$$

we will look for the distribution of births among the different generations 200 years after the birth of generation zero.

The generation with the most important contribution will be that of order $t / \mu_1 = 200 / 28.5 = 7.02$, that is, the seventh. Only the fifth, sixth, eighth, and ninth generations will also make considerable contributions.

Given that the generation of order zero comprised 100,000 newborn individuals, annual births at time $t = 200$ will be 10,487, of which 4 belong to the fifth generation, 1,275 to the sixth, 6,567 to the seventh, 2,481 to the eighth, and 164 to the ninth. Expressed as percentages these numbers correspond to 0.04, 12.15, 62.60, 23.65, and 1.56 percent of annual births at time $t = 200$.

With these results we conclude the study of the special case of the progeny of an original generation composed entirely of individuals born at the same time t. We return to the more general case of a population having any age distribution and progressing under the regime of constant mortality and fertility by age.

The Constants Q in the Expansion of $B(t)$. Let us return to the solution

$$B(t) = Q_1 e^{r_1 t} + Q_2 e^{r_2 t} + \cdots \tag{135}$$

of the integral equation

$$B(t) = \int_0^{\omega} B(t-a)\, p(a)\, m(a)\, da \tag{134}$$

To completely determine this solution it still remains to compute the coefficients Q, which depend on the initial conditions. We will see, in fact, that the constant Q_s corresponding to some root r_s of the fundamental equation

$$1 = \int_0^{\omega} e^{-ra}\, p(a)\, m(a)\, da \tag{137}$$

is linked to the function $B_1(t)$ representing the distribution in time of births in the first generation. In order to make the problem determinant, it is necessary that this function be given from time $t = 0$ until $t = a_2$ (the end of the reproductive period of the first generation) either directly or as a function of $B(t)$ following (199). This being understood, let us introduce the auxiliary constants

$$P_s = \int_0^{a_2} e^{-r_s t} B_1(t)\,dt \tag{214}$$

r_s being any root of equation (137).

Now, following (199)

$$B_1(t) = B(t) - \int_0^t B(t-a)\,\varphi(a)\,da \tag{199}$$

and in consequence

$$P_s = \int_0^{a_2} e^{-r_s t}\,dt \left\{ B(t) - \int_0^t B(t-a)\,\varphi(a)\,da \right\} \tag{215}$$

We have on the other hand

$$B(t) = \sum Q_j e^{r_j t} = Q_s e^{r_s t} + \sum_{-s} Q_j e^{r_j t} \tag{135}$$

the sum Σ comprising all of the terms formed with the roots r of the fundamental equation (137) and the symbol Σ_{-s} designating the same sum, omitting the term $Q_s\, e^{r_s t}$.

Introducing expression (135) in (215) we obtain

$$P_s = \int_0^{a_2} e^{-r_s t}\,dt \left\{ Q_s e^{r_s t} - \int_0^t Q_s e^{r_s (t-a)} \varphi(a)\,da \right\} +$$

$$\int_0^{a_2} e^{-r_s t}\,dt \left\{ \sum_{-s} Q_j e^{r_j t} - \int_0^t Q_j e^{r_j (t-a)} \varphi(a)\,da \right\}$$

$$= Q_s \int_0^{a_2} dt \left\{ 1 - \int_0^t e^{-r_s a} \varphi(a)\,da \right\} + R_s \tag{216}$$

Now,

$$\int_0^t e^{-r_s a}\varphi(a)\,da = \int_0^{a_2} e^{-r_s a}\varphi(a)\,da - \int_t^{a_2} e^{-r_s a}\varphi(a)\,da \qquad (217)$$

$$= 1 - \int_t^{a_2} e^{-r_s a}\varphi(a)\,da \qquad (218)$$

so that

$$P_s = Q_s \int_0^{a_2} dt \int_t^{a_2} e^{-r_s a}\varphi(a)\,da + R_s \qquad (219)$$

$$= Q_s \int_0^{a_2} e^{-r_s a}\varphi(a)\,da \int_0^a dt + R_s \qquad (220)$$

$$= Q_s \int_0^{a_2} a\,e^{-r_s a}\varphi(a)\,da + R_s \qquad (221)$$

Now, the expression R_s cancels since it is composed of terms of the form

$$R_{s,u} = \int_0^{a_2} e^{-r_s t}\,dt \left\{ Q_u e^{r_u t} - \int_0^t Q_u e^{r_u (t-a)}\varphi(a)\,da \right\} \qquad (222)$$

which, after a transform analogous to that of (216) to (219), become

$$R_{s,u} = Q_u \int_0^{a_2} e^{(r_u - r_s)t}\,dt \int_t^{a_2} e^{-r_u a}\varphi(a)\,da \qquad (223)$$

$$= Q_u \int_0^{a_2} e^{-r_u a}\varphi(a)\,da \int_0^a e^{(r_u - r_s)t}\,dt \qquad (224)$$

$$= \frac{Q_u}{r_u - r_s} \int_0^{a_2} e^{r_u a}\varphi(a)\,\{e^{(r_u - r_s)a} - 1\}\,da \qquad (225)$$

$$= \frac{Q_u}{r_u - r_s} \left\{ \int_0^{a_2} e^{-r_s a} \varphi(a)\, da - \int_0^{a_2} e^{-r_u a} \varphi(a)\, da \right\} \tag{226}$$

$$= 0 \tag{227}$$

since both r_s and r_u are roots of the fundamental equation (137). Hence, these terms are null, whatever the numbers s, u, provided that $s \neq u$. It follows that

$$R_s = 0 \tag{228}$$

and

$$P_s = Q_s \int_0^{a_2} a e^{-r_s a} \varphi(a)\, da \tag{229}$$

$$Q_s = P_s \Big/ \int_0^{a_2} a e^{-r_s a} \varphi(a)\, da \tag{230}$$

$$= \frac{\int_{a_1}^{a_2} B_1(t) e^{-r_s t}\, dt}{\int_0^{a_2} a e^{-r_s a} \varphi(a)\, da} \tag{231}$$

$$= \frac{\int_0^{a_2} e^{-r_s t}\, dt \left\{ B(t) - \int_0^t B(t-a) \varphi(a)\, da \right\}}{\int_0^{a_2} a e^{-r_s a} \varphi(a)\, da} \tag{232}$$

This result was obtained by introducing solution (135) in equation (215), which contains the values of $B(t)$ within the limits $t = 0$ and $t = a_2$; thus, the coefficients Q_s given by (232) depend on the values of $B(t)$ within those limits.

The Constants Q in Terms of the Cumulants μ of the Net Fertility Function $\varphi(a) = p(a)\, m(a)$. The denominator of (232) has a familiar aspect. In view of (154), (159), and (137) we can write

$$\frac{\int_0^{a_2} a e^{-r_s a} \varphi(\mathrm{a})\, \mathrm{da}}{\int_0^{a_2} e^{-r_s a} \varphi(a)\, da} = \mu_1 - \mu_2 r + \frac{1}{2}\mu_3 r^2 - \cdots \tag{233}$$

This expansion remains valid for complex values of r. In that case it is more convenient, however, to give the numerator of (232) the form

$$\int_0^{a_2} e^{-(u+iv)t} dt \left\{ B(t) - \int_0^t B(t-a)\varphi(a)\, da \right\} = U + iV \tag{234}$$

similarly, the denominator becomes

$$\mu_1 - \mu_2 (u+iv) + \frac{1}{2}\mu_3 (u+iv)^2 - \cdots = G - iH \tag{235}$$

so that Q_s can take two forms

$$Q_s = \frac{U+iV}{G-iH} = \frac{(UG-VH) + i(GV+HU)}{G^2+H^2} \tag{236}$$

and

$$Q_s' = \frac{(UG-VH) - i(GV+HU)}{G^2+H^2} \tag{237}$$

where from (235) the quantities G, H have the values

$$G = \mu_1 - \mu_2 u + \frac{\mu_3}{2!}(u^2 - v^2) - \frac{\mu_4}{3!}(u^3 - 3uv^2) + \cdots \tag{238}$$

$$H = \mu_2 v - \mu_3 uv + \frac{\mu_4}{3!}(3u^2 v - v^3) - \cdots \tag{239}$$

Introducing the constants Q_s, Q_s' in the solution (135) and putting

$$e^{(u+iv)t} = e^{ut}(\cos vt + i \sin vt) \tag{240}$$

we obtain, finally, the term

$$\frac{2e^{ut}}{G^2 + H^2}\{(GU - HV)\cos vt + (HU + GV)\sin vt\} \tag{241}$$

The constants U and V can be computed by numerical integration of the left part of (234), the values of the function $B(t)$ in the interval $0 < t < a_2$ being known.

Special Case: Descendants of an Initial Generation All of Whose Members Were Born at the Same Time. The numerator in the expression for Q_s takes a particularly simple form in the case which previously served us as an example, namely, in the case of a line of descent starting from a generation of order zero, all of whose N members were born at the same time, taken as zero. In that case the distribution with respect to time $B_1(t)$ of the "first" generation (comprised of daughters of generation zero) reduces simply to

$$B_1(t) = N\,p(t)\,m(t) = N\varphi(t) \tag{242}$$

and the numerator in the expression for Q_s becomes

$$N\int_0^{a_2} e^{-r_s t}\varphi(t)\,dt = N \tag{243}$$

r_s being a root of the fundamental equation (137).

In what has preceded we have supposed fertility $m(a)$ to be given, without pushing the analysis of this characteristic further. We can evidently decompose this general fertility $m(a)$ into two factors, of which one represents the proportion $\eta(a)$ of married women (spouses) among women of age a; and the other $m(a)$ represents marital specific fertility.[*] We can also distinguish legitimate and illegitimate fertility.[†] The introduction of frequency curves due to Karl Pearson for the representation of the function $\varphi(a) = p(a)\, m(a)$ was suggested by S. Wicksell[‡] and developed by A. J. Lotka.[§] Wicksell has also examined the relation between nuptiality and fertility,[**] and an application of the principles established by him to the statistics of the United States was made by Lotka.[††] For all of these developments the reader must be referred to the sources cited.

Finally, we also remark that if we have considered the function $m(a)$ to be independent of the proportions of the two sexes in the population, that is a legitimate approximation in the majority of cases, because this proportion does not vary seriously in general.[‡‡] In extreme cases, however, those variations can play a more important role.[§§]

---

[*] A. J. Lotka, "The size of American families in the eighteenth century," *Journal of the American Statistical Association*, 1927, p. 154; S. Wicksell, "Nuptiality, fertility and reproductivity," *Skandinavisk Aktuarietidskrift*, 1931, p. 125.

[†] S. Wicksell, *loc. cit.*

[‡] S. Wicksell, *loc. cit.*

[§] A. J. Lotka, "Industrial replacement," *Skandinavisk Aktuarietidskrift*, 1933, p. 51; *Annals of Mathematical Statistics*, 1939, p. 1.

[**] Wicksell, *loc. cit.* See also H. Westergaard, *Journal of the American Statistical Association*, 1920, p. 381.

[††] A. J. Lotka, *Annals of the Academy of Political and Social Science*, 1936, p. 1.

[‡‡] There naturally exists a relation between the proportion of the sexes at each age in the population, and the same proportion for newborns, this last being approximately constant. See on this subject A. J. Lotka, "A natural population norm," *Journal of the Washington Academy of Sciences*, 1913, p. 289; also Bortkiewicz, *Bulletin de l'Institut International de Statistique*, 1911, *v. 19(1)*, p. 63.

[§§] See A. J. Lotka, "The stability of the normal age distribution," *Proceedings of the National Academy of Sciences*, 1922, p. 343.

CHAPTER 5

Indices and Measures of Natural Increase

Relation between Annual Births at Two Periods Separated by the Mean Interval between Two Generations. In the case of a Malthusian population annual births increase or decrease according to the same "compound interest rate" r as the population, so that

$$\left.\begin{aligned} \frac{B(t)}{B(t-x)} &= e^{rx} \\ &= R_0^{x/T} \end{aligned}\right\} \tag{244}$$

this formula being valid for every value of x. In particular

$$\left.\begin{aligned} \frac{B(t)}{B(t-T)} &= \frac{B(t)}{B\left(t-m_1+\frac{m_2}{2!}r-\cdots\right)} \\ &= e^{rT} \\ &= R_0 \end{aligned}\right\} \tag{245}$$

or, as a first approximation,

$$\frac{B(t)}{B(t-\mu_1)} = R_0 \tag{246}$$

Formula (244) is valid for every value of x only in the case of a Malthusian population. By contrast, formulas (245) and (246) have much more general applicability. They represent a property of real populations on which a very practical approximate measure of reproductivity is based. To see this, let us return to relation (134), which also remains valid when mortality and fertility are functions of time t, so that we can write

$$B(t)=\int_0^\infty B(t-a)\,p(a,t)\,m(a,t)\,da \tag{134a}$$

Let

$$B(t-\bar{a})=\frac{\int_0^\infty B(t-a)\,p(a,t)\,m(a,t)\,da}{\int_0^\infty p(a,t)\,m(a,t)\,da} \tag{247}$$

the quantity $\bar{a}$ being in this sense a type of mean value of a, which in general will be a function of t.*

Dividing (134a) by

$$\int_0^\infty p(a,t)\,m(a,t)\,da$$

we obtain

$$\frac{B(t)}{\int_0^\infty p(a,t)\,m(a,t)\,da}=\frac{\int_0^\infty B(t-a)\,p(a,t)\,m(a,t)\,da}{\int_0^\infty p(a,t)\,m(a,t)\,da} \tag{248}$$

Hence,

$$\frac{B(t)}{R_0(t)}=B(t-\bar{a}) \tag{249}$$

---

* However, in practice the variation of $\bar{a}$ will remain within rather narrow limits.

Now, if $B(t - a)$ can be represented sufficiently well by a linear function within the limits $a_1 < a < a_2$,[*] the mean $\bar{a}$ defined by (247) will be essentially equal to the mean value α defined by the formula[†]

$$\alpha = \frac{\int_0^\infty a\, p(a,t)\, m(a,t)\, da}{\int_0^\infty p(a,t)\, m(a,t)\, da} = \mu_1(t) \tag{250}$$

We thus arrive at the relation

$$\frac{B(t)}{B(t-\mu_1)} = R_0 \tag{246}$$

If, rather than supposing that $B(t - a)$ has a linear form, we assume that from $B(t - a_2)$ to $B(t - a_1)$ this function will be approximately exponential, with a rate of increase ρ, we have within these limits

$$B(t-a) = B(t)\, e^{-\rho a} \qquad (a_1 < a < a_2) \tag{251}$$

and we find in an analogous manner

---

[*] It will even suffice that $B(t - a)$, considered as a function of a for a given value of t, be nearly linear within narrower limits, as the factor $m(a, t)$ has important values only in the neighborhood of $a = \mu_1$.

[†] See A. J. Lotka, *Journal of the American Statistical Association*, 1936, *v. 31*, the note at the bottom of p. 287. Given that $\varphi(x) = u + vx$, and $f(x)$ is some function of x, then the mean value $\varphi_m(x)$ will be equal to $\varphi(x_m)$, that is

$$\varphi_m(x) = \frac{\int \varphi(x) f(x) dx}{\int f(x) dx} = u + v \frac{\int x f(x) dx}{\int f(x) dx} = u + v x_m = \varphi(x_m)$$

$$\frac{B(t)}{\int_0^\infty p(a,t)\,m(a,t)\,da} = \frac{B(t)}{R_0(t)} = \frac{B(t)\int_0^\infty e^{-\rho a}\;p(a,t)\,m(a,t)\,da}{\int_0^\infty p(a,t)\,m(a,t)\,da}$$

$$= B(t)\,e^{-\rho\bar{a}} \tag{252}$$

Thus

$$e^{\rho\bar{a}} = R_0(t) \tag{253}$$

and the mean age $\bar{a}$ thus defined is nothing else than the mean interval T between two generations in a population growing at a rate of increase ρ. We have in consequence

$$\frac{B(t)}{B(t-T)} = R_0(t) \tag{245a}$$

On the other hand, the value of T in a population having a stable age distribution is, according to (160),

$$T_\rho = \mu_1 - \tfrac{1}{2}\mu_2\,\rho + \cdots \tag{160}$$

so that in such a case we can write more exactly

$$\frac{B(t)}{B\left(t - [\mu_1 - \tfrac{1}{2}\mu_2\,\rho + \cdots]\right)} = R_0(t) \tag{254}$$

In certain cases this formula corresponds better to observations than does formula (246).

These results merit recapitulation in the form of a proposition:

In any closed population the number R_0 *which expresses the ratio between total births in two successive generations, computed by starting from the actual mortality and fertility at the instant t, is approximately equal to the ratio between the annual number of births at the instant t and at the instant* $t - T$*, where* T *denotes the interval between the two successive generations.*

What is striking in this proposition is that in the course of a generation the rates of mortality and fertility can undergo considerable variation without the proposition losing its validity,* *provided that annual births* $B(t - a)$ *as a function of* a *do not deviate too greatly from a linear or exponential form in the neighborhood of* $a = \mu_1$. Several numerical examples will illustrate this rather remarkable fact.

Numerical Examples: First Example. The annual number of births registered in the United States during the years 1928 and 1929 was 2,330,000, including a correction for the 5.5 percent of the population who, at that time, still escaped birth registration. In 1900, that is to say a generation earlier ($\mu_1 = 28.3$ years), with the birth rate 29.5 per 1000,† and the population 75,995,000, the number of births should have been equal to 2,240,000. In this case, therefore, our formula informs us that

$$\frac{B(t)}{B(t - \mu_1)} = \frac{2.330}{2.240} = 1.040 \tag{255}$$

while direct computation using the rates of mortality and fertility observed in 1928 gives

$$R_0 = 1.049 \tag{256}$$

A similar computation for the years 1933 and 1905 gives

$$\frac{B(t)}{B(t - \mu_1)} = 0.883 \tag{257}$$

while the direct computation for 1933 shows that

---

* It is true that these changes will only very slightly affect the mean interval between two successive generations.

† According to an assessment made in 1925; see L. I. Dublin and A. J. Lotka, "The true rate of natural increase," *Journal of the American Statistical Association*, 1925, p. 317.

$$R_0 = 0.866 \tag{258}$$

Second Example. A particularly interesting example is presented by the distribution of annual numbers of births in the case of a logistic population having coefficients corresponding to the population of the United States. The results of the application of formulas (245) and (254) to this numerical series are presented in columns 7, 8 and 9 of Table 12. It can be seen that in the first part of the logistic series, corresponding to a nearly constant rate of increase, the figures obtained by formula (254) are in better accord with the results of direct computation of $R_0(t)$ than those found by formula (246).

It should be remarked that in general formula (244), which, for a Malthusian population, is valid for all x, would give a very inexact result if it were applied to any general population and to an arbitrary value of x. For example, in its application to the logistic population previously considered we would obtain in place of $\rho = 0.0105$ and $R_0 = 1.346$, the following values

t	x	$t-x$	ρ	$R_0(t)$
1920	20	1900	0.0912	1.295
1920	45	1875	0.1286	1.439

Replacement Index. The relation that we have just recognized between current births and those at a period a mean generation earlier could be employed to give us a measure at least close to net reproductivity R_0, and in consequence to the intrinsic rate of growth ρ. However, this would be an impractical method, since it requires fairly exact knowledge of the number of births separated by a considerable lapse of time, during which, in addition, the conditions on which our formulas are based could have varied.

However, a certain modification of our formulas will remedy this drawback.

We remark first that if $(\alpha_2 - \alpha_1)$ does not differ too greatly from T, we will have, again approximately,

Table 12. Ratio between annual births B at time t and births a mean generation T earlier in a logistic population, based on characteristics of the white female population of the United States, 1920

Year t	Rate of natural increase r	Mean generation length T	Annual births $B(t)$	Annual births $B(t-\mu_1)$	Annual births $B(t-T)$
1	2	3	4	5	6
1800	0.0306	27.78	218,697	91,897	93,870
1825	0.0296	27.80	454,475	197,055	201,064
1850	0.0277	27.84	890,975	411,709	419,184
1875	0.0243	27.92	1,567,415	815,688	827,261
1900	0.0191	28.04	2,330,669	1,462,024	1,474,927
1920	0.0143	28.15	2,797,246	2,082,021	2,091,788
1925	0.0131	28.17	2,882,723	2,231,455	2,240,209
1950	0.0077	28.30	3,164,261	2,824,622	2,827,609
1975	0.0040	28.38	3,303,823	3,136,262	3,137,024
2000	0.0020	28.43	3,373,610	3,289,582	3,289,754

Year t	Net reproductivity calculated according to $\frac{B(t)}{B(t-\mu_1)}$	Net reproductivity calculated according to $\frac{B(t)}{B(t-T)}$	Net reproductivity calculated according to $R_0(t)$	Intrinsic rate of natural increase, based on the values of net reproductivity in columns 7, 8, 9: from (7)	from (8)	from (9)
1	7	8	9	10	11	12
1800	2.380	2.330	2.329	0.0312	0.0304	0.0304
1825	2.306	2.260	2.260	0.0301	0.0293	0.0293
1850	2.164	2.125	2.127	0.0277	0.0271	0.0271
1875	1.922	1.895	1.899	0.0234	0.0229	0.0229
1900	1.594	1.580	1.588	0.0166	0.0163	0.0165
1920	1.344	1.337	1.346	0.0105	0.0103	0.0105
1925	1.292	1.287	1.296	0.0091	0.0089	0.0092
1950	1.120	1.119	1.125	0.0040	0.0040	0.0042
1975	1.053	1.053	1.056	0.0018	0.0018	0.0019
2000	1.026	1.025	1.027	0.0009	0.0009	0.0009

Explanation according to column number:

(2) From the formula $r = \frac{0.0314}{1+e^{0.0314(t-1914.1)}}$. (3) $T = \mu_1 - \frac{1}{2}\mu_2 r$.

(4, 5, 6) $B(t)$, $B(t-\mu_1)$, $B(t-T)$ from text formula (109) or (126).

(9) Calculated by multiplying $R_0(1920)$ by the ratio of $B(t)$ to annual births $\Sigma\, m(a, 1920)\, c(a, t)$ which would occur at time t under the fertility regime of 1920.

(10, 11, 12) Calculated from the formula $\rho(t) = \frac{1}{\mu_2}\left\{ \mu_1 - \sqrt{\mu_1^2 - 2\mu_2 \ln R_0(t)} \right\}$.

Note: The estimates for the three final dates 1950, 1975, and 2000 are given only to complete the example, and should not be taken as forecasts.

$$\frac{B(t)}{B(t-[\alpha_2-\alpha_1])}=e^{\rho(\alpha_2-\alpha_1)} \tag{259}$$

that is to say,

$$\frac{B(t-\alpha_1)}{B(t-\alpha_2)}=e^{\rho(\alpha_2-\alpha_1)} \tag{260}$$

$$=R_0^{(\alpha_2-\alpha_1)/\alpha} \tag{261}$$

On this relation rests a very practical measure of net reproductivity introduced by W. S. Thompson* and used extensively by F. Lorimer and F. Osborn (1934) in their book *The Dynamics of Population Growth*. This measure, called the *Replacement Index*, is constituted, first, by forming the quotient: (number of children ages 0 to 5 years) divided by (number of women ages 15 to 45) in the population at time t; and second, the same quotient for a stationary population with an age distribution corresponding to the life table applicable at time t; third, upon dividing the first quotient by the second. The age limits of the two groups — lower and upper, children and women — are, moreover, arbitrary up to a certain point, provided they are suitably chosen.

In analytical language this definition of the replacement index has the form

$$J=\frac{\int_p^q c(a)\,da}{\int_u^v c(a)\,da}\div\frac{\int_p^q p(a)\,da}{\int_u^v p(a)\,da} \tag{262}$$

the functions $c(a)$, $p(a)$ being with respect to the female population, the symbols p, q signifying the age limits of the group of young girls, the symbols u, v the age limits of the groups of women.

Changing the order of the factors, we obtain

---

* *The Ratio of Children to Women*, published by the Bureau of the Census, Washington, 1920.

$$J = \frac{\int_p^q c(a)\,da}{\int_p^q p(a)\,da} \div \frac{\int_u^v c(a)\,da}{\int_u^v p(a)\,da} \tag{263}$$

$$= \frac{\int_p^q B(t-a)\,p(a)\,da}{\int_p^q p(a)\,da} \div \frac{\int_u^v B(t-a)\,p(a)\,da}{\int_u^v p(a)\,da} \tag{264}$$

$$= \frac{B(t-\alpha_1)}{B(t-\alpha_2)} \tag{265}$$

$$= R_0^{(\alpha_2-\alpha_1)/\alpha} \tag{266}$$

taking into account (261), always on condition that the age limits p, q; u, v of the lower and upper group are suitably chosen, and that the difference $\alpha_2 - \alpha_1$ does not differ too greatly from α.

We remark also that the relation (266) between the index J of replacement and net reproductivity R_0 is nearly independent of the specific life table employed in the computations. That derives from the fact that the mean ages α_1 of the lower group and α_2 of the upper group, as well as the mean age α of the net fertility curve, are only slightly influenced by a change of life tables.

Application. In Figure 16 the replacement indices J for 45 of the 48 United States in 1930, computed according to formula (262), are represented* by the abscissas in a rectangular system whose ordinates represent the corresponding values of R_0. The age limits employed were 0 to 5 years for the children (girls) and 15 to 45 for the women, and the corresponding values for the mean ages of these groups were $\alpha_1 = 2.48$, $\alpha_2 = 32.29$. The solid curve drawn through the series of points representing

* For the details, the reader can consult the article already cited, *Journal of the American Statistical Association*, 1936, p. 273.

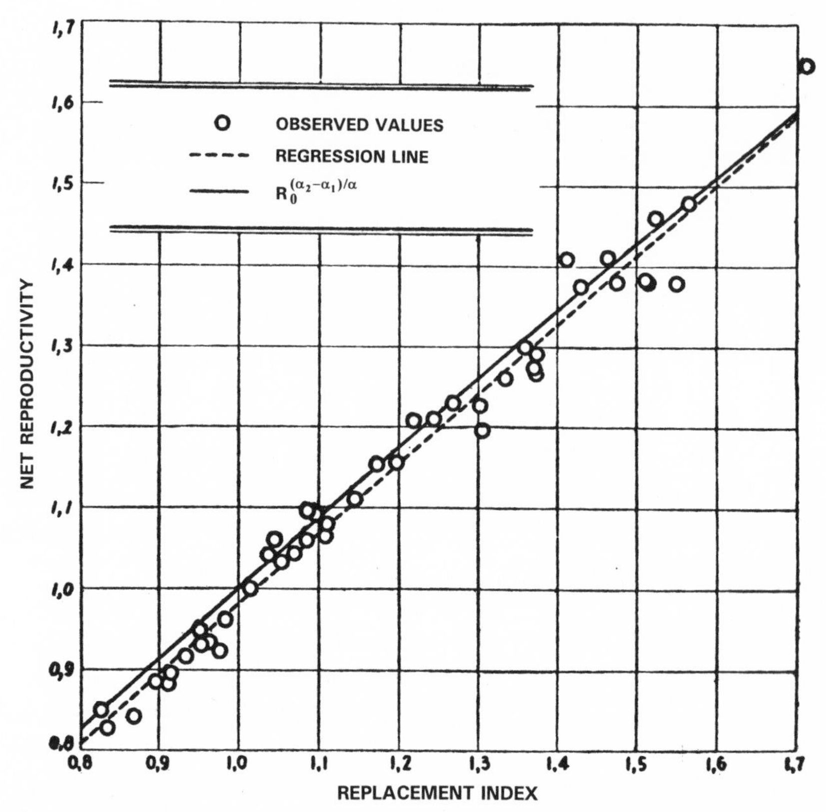

Figure 16. Relation between the replacement index J and net reproductivity R_0.

the 45 states is the trace of the function $R_0^{(\alpha_2-\alpha_1)/\alpha}$. One sees that this curve represents the series of points very well, despite the fact that the curve is computed with the aid of the life table for the entire United States, while mortality in certain states differs very sensibly from the mean for the entire country. This provides an illustration of the fact noted earlier, that relation (266) is nearly independent of the specific life table employed in the computations.

Variation of the Replacement Index according to the Choices of Lower and Upper Age Groups. We have remarked that the choice of age limits for the lower and upper groups is arbitrary up to a certain point, provided that the ratio $(\alpha_2 - \alpha_1) / \alpha$ does not differ too greatly from unity. We can therefore ask what will be the effect of varying the choice of the age limits.

Variation of the Lower Age Group. The results of computations executed by varying the age limits of the lower group of ages are assembled in Table 13. By way of example, a population having a stable age distribution was chosen, assigning six different values to the rate of increase, and taking as a base the 1929–1931 American life table.

The table shows that the choice of the upper age limit for the lower age group only slightly influences the replacement index computed on a fixed base (ages 20 to 45) for the upper group, but varying the upper limit of the lower age group from 1 year to 5 years.

Parallel Variation of the Lower and Upper Age Groups. The application of formula (266) requires that the difference $\alpha_2 - \alpha_1$ not diverge too greatly from the value of α; apart from that the choice of groups for which α_2, α_1 are the mean ages remains arbitrary.

In place, then, of combining, for example, the upper group 20 to 45

Table 13. Variation in the replacement index *J* corresponding to variations in the upper age limit of the lower age group for a population having a stable age distribution. Computations based on the United States white female life table for 1929–1931

Lower group	Intrinsic rate of natural increase per capita					
	0.03640	0.02800	0.01736	0.00265	–0.02170	–0.04516
	Replacement index					
Age 0*	3.130	2.420	1.738	1.089	0.490	0.221
Less than 1 year	3.052	2.373	1.717	1.087	0.498	0.228
Less than 2 years	3.004	2.344	1.704	1.086	0.503	0.233
Less than 3 years	2.995	2.315	1.690	1.084	0.508	0.237
Less than 4 years	2.905	2.285	1.677	1.083	0.513	0.243
Less than 5 years	2.856	2.255	1.663	1.082	0.518	0.248
30 to 35 years**	2.752			1.079		

* Annual births

** The upper age group is taken between the limits 20 to 45 years of age, except in the final line of the table, where it is taken as 50 to 75 years.

years of age with the lower group 0 to 5 years, to derive the value of J or R_0 at time $t - \alpha_1 = t - 2.5$ (approximately), we can combine the groups 25 to 50 years and 5 to 10 years; 30 to 55 years and 10 to 15 years, and so forth, to extract the values of J or R_0 for a corresponding period, namely, at about time $t - 7.5$; $t - 12.5$; etc. In this manner we can obtain a suggestion of the evolution of these indices over time. A computation made on the basis of the white female population of the United States using the 1930 census, and on the basis of the 1929–1931 life table, gives the results in Table 14.

Table 14. Trend in the replacement index computed for successive age groups. Based on data for the United States, white females, 1929–1931

Lower group	Upper group	Replacement index	Average year of birth of lower group
1	2	3	4
0–5	20–45	1.088	1928
5–10	25–50	1.309	1923
10–15	30–55	1.359	1918
15–20	35–60	1.444	1913
20–25	40–65	1.558	1908
25–30	45–70	1.595	1903
30–35	50–75	1.700	1898
35–40	55–80	1.933	1893
40–45	60–85	1.886	1888
45–50	65–90	1.878	1883

One sees there the decrease in the replacement index J. It is true that only a qualitative value can be attributed to this information, for the United States having formerly received large numbers of immigrants, the individuals comprising the lower group do not represent exclusively the children of the upper group, as the exact application of the formula would require.*

Excess of the Crude Rate of Increase over the Intrinsic Rate in a Logistic Population. The two series, that of the crude rate of increase

* In its application to the group 0 to 5 years this objection is without value, since the arrival of such young children without their parents is a very rare event.

and that of the intrinsic rate, computed for a population expanding according to the logistic model, present a phenomenon that requires our attention. We see, in columns 2 and 12 of Table 12, and in the corresponding graph (Figure 17), that everywhere except at $t = \pm \infty$ the intrinsic rate is less than the crude rate. Thus, the fact that the intrinsic rate of the United States is below the crude rate cannot be considered abnormal. The two rates are only equal for a Malthusian population or a stationary population. Now, Malthusian increase cannot possibly persist indefinitely, and the stationary state has not yet been attained in the majority of countries today. In consequence, the lower value of the intrinsic rate relative to the crude rate is not disturbing in itself. It is true that when we see it taking negative values we have good reason to ask ourselves where that will lead us.

Advantages and Disadvantages of the Various Indices or Measures of Natural Increase. We have presented several different indices of the natural increase of a population.

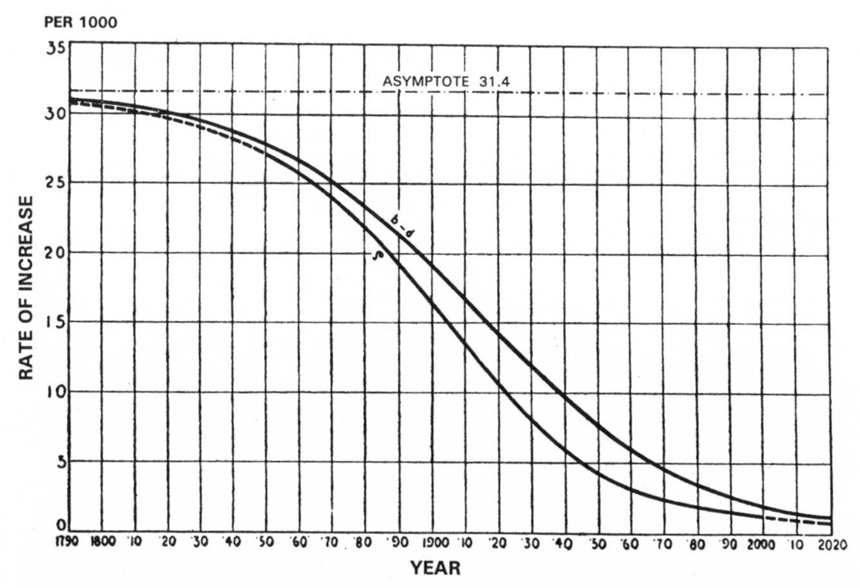

Figure 17. Comparison of the crude natural increase rate ($b - d = r$) with the intrinsic rate ρ in a logistic population. Based on the characteristics of the population of the United States.

1. The most common is the crude rate r, the excess of the birth rate over the death rate. Its advantage consists in the facility of its computation whenever demographic statistics exist. Its defect is its dependence on the age distribution of the population, which sometimes renders its information very misleading.

2. Net reproductivity, R_0, introduced by Böckh,* has greater merit, for it gives a measure essentially independent of the age distribution of the population. Its computation, it is true, requires knowledge of fertility and mortality by age. However, experience shows that it suffices if fertility is given by age groups in five year intervals. On the other hand, this measure is influenced by the mean interval between generations, and for this reason R_0 computed for the male part of the population is higher than that which the female population gives. As well, the diagnostic value of R_0, which has been much vaunted by certain authors, is limited. It is true that where the crude growth rate r is still positive, whereas $R_0 - 1$ has become negative, this warns us that the information from the crude rate is misleading. But in the case where r is positive, and $R_0 - 1$ is as well, as in the case of the United States in 1920, net reproductivity would in no way have warned us that the crude rate gave us an exaggerated idea of the intrinsic capacity of growth of the population. It is because the value of $(R_0 - 1)$ has become negative in many countries

---

* *Translators' note:* The net reproduction rate was introduced, or nearly introduced, on two earlier occasions by William Farr. From the Census Report of 1851: "... if the population increases at the rate of 1.329 annually, and if the intervening time from generation to generation is 33½ years, it follows that the increase from generation to generation is 55 per cent.; or that every 1,000 women are succeeded, at the interval of 33½ years, by 1,553 women ..." This is Lotka's expression (245), but Farr did not have a means for estimating the generation length and probably did not see that it had mathematical significance. In 1880, using Norwegian data, he computed the number of children who would be born to mothers at each age in a life table population, adjusting the l_x series for the sex ratio at birth. Had he summed the birth column of his table it would have given him the net reproduction rate directly, but he apparently did not appreciate its significance, and did not notice that it related to his earlier start. See Farr, *Vital Statistics: A Memorial Volume of Selections from the Reports and Writings of William Farr*, The Scarecrow Press, Metuchen NJ, 1975, p. 28 for a reprint of his 1851 remark; and Farr, "English reproduction table," *Philosophical Transactions*, 1880, Part 1: 281–288 for his table. A comment on Farr's 1880 paper will be found in F. M. M. Lewes, "A note on the origin of the net reproduction ratio," *Population Studies,* 1984, *v. 38*, pp. 321–324.

today that this imperfection in R_0 has escaped certain authors.

3. We also mention briefly a measure proposed by G. Knibbs,[*] namely the product of the birth rate and life expectancy. I have shown elsewhere[†] that the product $b\ L_0$ gives a value approaching R_0 in the case of a population having a stable age distribution, but in general $b\ L_0$ is not, even approximately, equal to this quantity.

4. Similarly it suffices to note in passing the idea of corrected rates, as would be obtained by applying the values of fertility and mortality by age to a population having an age distribution proportional to the l_x values in the life table. The computation of rates corrected in this manner rests on contradictory bases, since, unless the population is in fact stationary, the age distribution introduced in the computations is in contradiction with the given fertility and mortality.

5. The replacement index *J*, as we have seen, actually gives a value approaching (and generally very near) the net reproductivity R_0. It has the advantage of requiring only a knowledge of the age distribution of the population. It is not necessary to know either fertility or mortality.

As an approximate value for R_0 the index *J* naturally shares its advantages and disadvantages. It has as well the defect that it is not an entirely determinant measure, for its value depends up to a certain point on the age limits chosen for the lower and upper age groups (daughters and mothers). Nonetheless, it should be acknowledged that the replacement index is a very ingenious tool.

6. The intrinsic rate of increase, ρ, has many advantages. It gives us all the information of net reproductivity, which, in any event, is only a sub-product of the computation of ρ. But it tells us much more. In the case where $R_0 - 1$ is positive, and where R_0 alone tells us nothing to put us on guard against the superficial indications of the crude rate, the intrinsic rate gives us precisely the criterion and the measure that should replace the crude rate. In addition, ρ is expressed directly in familiar units; it is an annual rate *per capita*, and hence can be compared directly to the crude rate and all other annual rates.

---

[*] G. H. Knibbs, *Mathematical Theory of Population*, 1917, p. 294.

[†] A. J. Lotka, "The measure of net fertility," *Journal of the Washington Academy of Sciences*, 1925, p. 469.

The computation of ρ is also very simple. After having computed the moment of order zero R_0 of the function $p(a)\, m(a)$, nothing more remains except to compute the moments of the first and second order, an affair of ten minutes with a modern machine, and then there is a simple quadratic equation to resolve, that is all. Certain authors* have given an impression that there is something difficult in that; in fact it is excessively simple.

But the greatest advantages of ρ manifest themselves when we develop demographic analysis further. It is then seen that the intrinsic rate of natural increase plays an important role in widely varied problems. In particular, introduced in the formulas for the birth rate and death rate for a Malthusian population, it permits computation of the intrinsic rates of birth and death corresponding to the given functions of fertility and mortality, about which net reproductivity R_0 has in no way informed us.

* See for example G. Paulinus, "Prolegomena zu einer Bevölkerungsprognose," *Inaugural-Dissertation*, Leipzig, 1934.

CHAPTER 6

Relations Involving Fertility by Birth Order

Relations Bearing on the Phenomenon of the Family. The higher they ascend in the evolutionary scale, the more biological organisms depend, during their early growth, on protection and apprenticeship from their parents, or at least from their mother. In the human species that phenomenon reaches its highest degree and gives rise to the family, an institution whose fundamental influence extends over all phases of the life of the individual, of the community, and of nations.

In what follows we will therefore examine some relations that bear on the composition of the family in a closed population existing under the regime of a fixed mortality and fertility table. To simplify the discussion we will take into account only legitimate children and only first marriages.

Proportion Married and Marriage Rate. Between the proportion $s(a)$ of unmarried women, or married women $1 - s(a) = H(a)$, at age a, and the rate of marriage $\nu(a)$, in the ideal case where all marriages are first marriages, there evidently exists the simple relation*

$$\frac{d\,s(a)}{da} = -\nu(a)\,s(a) \qquad (267)^{\dagger}$$

* This relation has been noted by, among others, H. Westergaard, *Journal of the American Statistical Association*, 1920, p. 381; and by S. Wicksell, *Skandinavisk Aktuarietidskrift*, 1931, p. 125.

† *Translators' note:* In expressions (267) *et seq*. Lotka uses the notation: E, ε = enfants (children), e = épouse (wife), m = mère (mother) , δ = divorcées, ζ = stérilité, η = épouses effectives (women currently married), υ = veuves (widows).

$$s(a) = e^{-\int_0^a \nu(a)\,da} \tag{268}$$

The enumeration of a population by age and marital status gives us the values of the ratio $s(a)$ directly; the values of the rate of marriage ν are found immediately by numerical differentiation using formula (267).

Number of Children per Wife. Let us fix our attention on a cohort of N newborn daughters, and follow their career up to the point that the last survivor is carried away by death. At age a there are $N\,p(a)$ survivors, of whom $N\,p(\mathrm{a})\,s(a)$ are unmarried; among these there will be $N\,p'(a)\,s(a)\,da$ decedents before attaining age $a + da$. Ultimately, when the entire cohort is defunct, the number of women dying celibate will be expressed by the integral

$$S = N \int_0^{\omega} p'(a)\,s(a)\,da \tag{269}$$

and the number of women who had been married during their lives will be equal to the difference

$$N - S = N\left\{1 - \int_0^{\omega} p'(a)\,s(a)\,da\right\} \tag{270}$$

On the other hand, the number of children brought into the world by the cohort will be $N \int_0^{\omega} p(a)\,m(a)\,da$, so that the number of children* per wife will be equal to

$$E_e = \frac{\int_0^{\omega} p(a)\,m(a)\,da}{1 - \int_0^{\omega} p'(a)\,s(a)\,da} \tag{271}$$

---

* Here the symbol $m(a)$ refers to children of both sexes.

Number of Children per Mother. The statistics of several countries give information not only on overall fertility $m(a)$, but also on fertility specific to birth order. If, therefore, we designate by $m_1(a)$ the number of first births *per capita* among women of age a, then the number of women in the cohort N who will have a first child, who, therefore, will become mothers, will be $N\int_0^{\omega} p(a)\, m_1(a)\, da$. In consequence, the number of children per mother will be

$$E_m = \frac{\int_0^{\omega} p(a)\, m(a)\, da}{\int_0^{\omega} p(a)\, m_1(a)\, da} \tag{272}$$

Rate of Sterility. If the number of children per mother is greater than that of children per wife, it is because a certain proportion of marriages remain sterile. This proportion, this "rate of sterility," can be expressed by the formula

$$\zeta = 1 - \frac{E_e}{E_m} = \frac{E_m - E_e}{E_m} \tag{273}$$

or in a more direct fashion*

$$\zeta = 1 - \frac{\text{number of women in the cohort who become mothers}}{\text{number of women in the cohort who become wives}}$$

so that

$$\zeta = 1 - \frac{\int_0^{\infty} p(a)\, m_1(a)\, da}{1 - \int_0^{\infty} p'(a)\, s(a)\, da} \tag{274}$$

* Formula (274) gives the complete measure of sterility independent of cause: physiological sterility, premature death of one or the other of the two spouses, marriage after the limiting age of reproduction, etc.

Relation between the Mean Number of Children per Family and the Rate of Increase. It is evident that there exists a more or less close relation between the mean number of children per family and the rate of increase of a closed population. For the case of a stable population this relation takes a definite form. In this case the individuals in a cohort of N newborn daughters give birth, in the course of their lives, to $R_0 N$ daughters, therefore to 2.06 $R_0 N$ children (see p. 166). Let H be the proportion of individuals in the cohort who become wives, and M the proportion who become mothers. Then the number of children per wife will be

$$E_e = \frac{2.06\,R_0}{H} \tag{275}$$

so that

$$R_0 = \frac{H\,E_e}{2.06} = e^{\rho T} \tag{276}$$

$$\rho = \frac{1}{T}\ln\frac{H\,E_e}{2.06} \tag{277}$$

In an analogous manner we obtain*

$$\rho = \frac{1}{T}\ln\frac{ME_m}{2.06} \tag{277a}$$

Application. The following numerical example, taken from data for the United States population (white race) in 1930, will serve not only to illustrate the use of the formulas in the preceding paragraphs, but also to show the importance of having sufficiently accurate data.

We find, according to formulas (271) and (272)

* See A. J. Lotka, "The size of American families in the eighteenth century," *Journal of the American Statistical Association*, 1927, pp. 161–166.

$$\left.\begin{aligned} E_e &= \frac{2.046}{0.8308} = 2.46 \\ E_m &= \frac{2.046}{0.6290} = 3.25 \end{aligned}\right\} \tag{278}$$

so that, from (273)

$$\zeta = 1 - \frac{E_e}{E_m} = 1 - \frac{2.46}{3.25} = 0.243 \tag{279}$$

We would thus have a crude level of sterility of 24.3 percent, a frightening number.

However, the registration of births in several of the States is still very imperfect. We have estimated that perhaps 8 percent of births escape being reported. Introducing that correction we find

$$\left.\begin{aligned} E_e &= \frac{2.210}{0.8308} = 2.66 \\ E_m &= \frac{2.210}{0.6793} = 3.25 \end{aligned}\right\} \tag{280}$$

$$\zeta = 0.182 \tag{281}$$

that is to say, a level of sterility of 18.2 percent, a percentage still very high, but certainly much more moderate than 24.3.

It should be noted that in these computations no correction has been made for illegitimate births, since we lack statistical data on this subject, taking birth order into account.

Composition of the "Stationary" Family. Under the regime of the fertility and mortality of the United States in 1930, when the intrinsic rate of increase was 0.003 *per capita*, we have computed a mean family of 2.66 children per wife and 3.25 children per mother. One might wonder what will be the family composition sufficient to exactly maintain the equilibrium between fertility and mortality, so that the rate of increase becomes zero. It will be necessary, evidently, that the cohort of N

newborn daughters, of whom NH become wives and $NH\frac{E_e}{E_m}$ become mothers, gives birth to N daughters. Therefore the number of daughters per wife must necessarily be

$$\frac{N}{NH}=\frac{1}{H} \tag{282}$$

and the number of daughters per mother

$$\frac{N\,E_m}{N\,H\,E_e}=\frac{1}{H(1-\zeta)} \tag{283}$$

These formulas give the number of daughters per wife or per mother that would be necessary to maintain the demographic equilibrium. The corresponding number of children is obtained by multiplying by the ratio of the number of total births to the number of female births; that ratio, relatively constant in the civilized countries, is about equal to 2.06.

In the example* cited above we find, thus, for the stationary family in 1930,

$$\left.\begin{aligned} E_e' &= \frac{2.057}{0.8308} = 2.47 \text{ children per wife} \\ E_m' &= \frac{2.057}{0.8308\,(1-0.182)} = 3.03 \text{ children per mother} \end{aligned}\right\} \tag{284}$$

Mean Completed Family Size. It should be remarked that the numbers obtained in this fashion represent the mean *completed* family, in the sense that every woman in the cohort is followed from her birth to her death when we form the integrals of formula (272).†

---

* In the example we have ignored the influence of illegitimate births. In view of the imperfect registration system, the numbers cited will be sufficiently realistic to serve by way of example.

† For a detailed discussion of this subject the reader should see the articles by C. Gini

Intact Completed Families. It is particularly interesting to study the composition of the family under conditions entirely favorable to its complete development, that is, in the case where both spouses remain alive, without dissolution of their marriage, at least up to the end of the woman's reproductive period.

Let $\eta(a)$ be the proportion of women at age a who are currently married (wives); also, let $\upsilon(a)$ be the proportion widowed, and $\delta(a)$ the proportion divorced. Then, $m(a)$ being the overall reproductivity of women of age a (whatever their marital status), it is evident that the reproductivity of married women will be

$$m'(a) = \frac{m(a)}{\eta(a)} \tag{285}$$

Let us fix our attention on a cohort of N women, of whom $N\,\eta(a)$ are currently married as of age a, and pass through the reproductive period without any of them being carried away by death; in the course of their lives these women will have a number ε of children

$$\varepsilon = N \int_0^{\infty} \eta(a)\, m'(a)\, da \tag{286}$$

However, in an actual population a certain proportion $\upsilon(a)$ of these women find themselves widowed by age a, and an additional proportion $\delta(a)$ are divorced by age a. Let us choose our cohort in such a way that it contains only women who have not suffered either of these misfortunes, or, what amounts to the same thing, let us suppose that deceased husbands are immediately replaced and divorced couples immediately remarry; under these hypotheses the proportion of women currently married as of age a will be increased from $\eta(a)$ to

$$\eta'(a) = \eta(a) + \upsilon(a) + \delta(a) \tag{287}$$

and the number of children born to this cohort will be

---

and F. Savorgnan in the *Bulletin de l'Institut International de Statistique*, 1934, *v. 27*, p. 40, and the *Revue* of the same Institute, 1933, *v. 1*, p. 23.

$$\varepsilon' = N \int_0^\infty \eta'(a)\, m'(a)\, da \tag{288}$$

On the other hand, the number of mothers in the cohort will evidently be equal to the number of "first" children born under the same conditions, that is, to

$$\varepsilon_1' = N \int_0^\infty \eta'(a)\, m_1'(a)\, da \tag{289}$$

if we allow the factor $\eta'(a)$ to apply both to the probability of having a first child and to the probability of having a child of any order.

The number of children per mother in the cohort will thus be

$$E_m' = \frac{\varepsilon'}{\varepsilon_1'} = \frac{\int_0^\infty \eta'(a)\, m'(a)\, da}{\int_0^\infty \eta'(a)\, m_1'(a)\, da} \tag{290}$$

Application. Utilizing 1930 data for the United States, from (280) we obtain a mean completed family size of 3.253 children per mother, and from (290) a mean intact completed family size of 3.434 children per mother. It is thus seen that under modern conditions intact completed families only marginally exceed all completed families.

CHAPTER 7

Relations Involving the Survival Functions of Two Individuals

Dissolved Families. The dissolution of families resulting from the death of one or the other of the parents, or of both, is a sociological and economic problem of the first importance. There is therefore a practical interest in computing the frequency of this demographic phenomenon in its relation to the biometric characteristics of a population

Proportion of Orphans at a Given Age *a*. Female Maternal Orphans [Daughters Who Have Lost Their Mother]. Let $B(t)$ be the annual number of female births at time t; $m(n)$ the annual number of such births per woman of age n; and $p(n)$ the proportion of newborn females who attain the age n. Then the annual number of female births which arise from mothers at ages comprised between n and $n + dn$ will be

$$B(t-n)\,p(n)\,m(n)\,dn \tag{291}$$

and that number will be a fraction

$$\frac{B(t-n)\,p(n)\,m(n)\,dn}{B(t)} \tag{292}$$

of total female births at time t. Let us apply this reasoning to daughters, numbering $B(t-a)\,p(a)\,da$, who at time t are at ages comprised between a and $a + da$. Evidently the proportion of these daughters arising from mothers at ages n to $n + dn$ at their births will be

$$\frac{B(t-a-n)}{B(t-a)}\,p(n)m(n)dn \tag{293}$$

and their number will be

$$B(t-a-n)\,p(n)\,m(n)\,dn\;p(a)\,da \tag{294}$$

The age of these daughters being a years, that of their mothers will be $(a + n)$ years, so that the number of these mothers who are living at time t will be

$$B(t-a-n)\,p(a+n)\,m(n)\,dn\;p(a)\,da \tag{295}$$

Finally, the total number of daughters of age a to $a + da$ at time t who still have a living mother will be

$$p(a)\,da\int_0^{\omega} B(t-a-n)\,p(a+n)\,m(n)\,dn \tag{296}$$

and the proportion of these daughters among all daughters of the same age will be

$$\int_0^{\omega} \frac{B(t-a-n)}{B(t-a)}\,p(a+n)\,m(n)\,dn$$

$$= \int_0^{\omega} \frac{p(a+n)}{p(n)}\,\frac{B(t-a-n)}{B(t-a)}\,p(n)\,m(n)\,dn \tag{297}$$

$$= \frac{p(a+\bar{n})}{p(\bar{n})}\,\frac{B(t-a-\bar{n})}{B(t-a)}\int_{a_1}^{a_2} p(n)\,m(n)\,dn \tag{298}$$

Now, if the product

$$\frac{p(a+n)}{p(n)}\,\frac{B(t-a-n)}{B(t-a)}$$

considered as a function of n, diverges only slightly from a linear form between the limits a_1 and a_2,* the mean $\bar{n}$ in the left part of (298) does not differ much from the mean defined by the formula

$$\bar{n} = \int_{a_1}^{a_2} n\, p(n)\, m(n)\, dn \Big/ \int_{a_1}^{a_2} p(n)\, m(n)\, dn \tag{299}$$

that is to say, from the mean age of the fertility curve. But in that case relation (249) gives

$$\frac{B(t-a-\bar{n})}{B(t-a)} = \frac{1}{R_0} \tag{300}$$

We have on the other hand,

$$\int_{a_1}^{a_2} p(n)\, m(n)\, dn = R_0 \tag{301}$$

The proportion of daughters of age a whose mothers are still living thus reduces simply to the ratio

$$\frac{p(a+\bar{n})}{p(\bar{n})} \tag{302}$$

and, finally, the proportion of daughters of age a who are maternal orphans will be

$$\Omega(a) = 1 - \frac{p(a+\bar{n})}{p(\bar{n})} \tag{303}$$

Male Maternal Orphans [Sons Who Have Lost Their Mother]. Formula (303) no longer contains any characteristic appropriate exclusively to female children, since the mean age of mothers at the birth of

---

* It even suffices that this linearity occurs within narrower limits, since the function $p(n)\, m(n)$ has a considerable value only in the region of its maximum.

daughters does not differ sensibly from their mean age at the birth of sons. Formula (303) thus applies equally to the computation of the proportion of sons of age a who have lost their mother, that is, who are male maternal orphans.

Male Paternal Orphans [Sons Who Have Lost Their Father]. The reasoning by which we have obtained formula (303) for the proportion of male maternal orphans evidently applies as well to the computation of male paternal orphans, provided that the survival function $p(a)$, which in this case is with respect to fathers, is taken from male mortality data.

Children Orphaned from Their Birth. A correction remains to be made in formula (303) in its application to maternal as well as paternal orphans.

As for the former, a systematic error is introduced in applying to women at the moment of delivery the survival function taken from data for the entire female population; this error is all the more important as the special risks of delivery concentrate precisely at the time of the birth of the children for whom we seek the proportion having lost their mother. In fact, for these mothers mortality does not have the incremental appearance of survival curves, but, on the contrary, passes through a pronounced maximum at the time of each birth. In consequence a certain number of children find themselves orphans from their birth.

In a somewhat analogous manner, a certain proportion of children are born paternal orphans, since their fathers are exposed during nine months after the conception and before the birth of these children to the normal risks of death.

The correction to apply to the computation due to this second circumstance is evidently very simple. It suffices to introduce in formula (303) the father's age nine months before the birth of the child, in place of his age at the birth itself. Hence, our formula becomes

$$\Omega'(a) = 1 - \frac{p(a+\bar{n})}{p(\bar{n} - \frac{3}{4})} \tag{304}$$

At the moment of the child's birth we have

$$a = 0 \tag{305}$$

$$\Omega_0{}'(0) = 1 - \frac{p(\bar{n})}{p(\bar{n} - \frac{3}{4})} \tag{306}$$

$$= \frac{p(\bar{n} - \frac{3}{4}) - p(\bar{n})}{p(\bar{n} - \frac{3}{4})} \tag{307}$$

$$= -\frac{3}{4} \frac{d\, p(\bar{n})}{p(\bar{n})\, dn} \tag{308}$$

The proportion of posthumous children among total births is thus equal to three-fourths of the mortality rate at the mean age $\bar{n}$ of fathers at the children's birth. According to the statistics for the United States in 1920, the value of $\bar{n}$ for fathers was 32.1, and the value of the mortality rate at that age was 0.0069, so that the proportion of posthumous children should have been about $(3 / 4) \times 0.0069 = 0.0052$ or 5.2 per thousand. This figure is probably a little too high, since males who become fathers probably have a level of mortality a little below the mean. A more recent life table would also give a less elevated figure. For example, using the mortality of the United States in 1930 we find a proportion $(3 / 4) \times 0.0042 = 0.0031$ or 3.1 per thousand of posthumous children.

The correction to apply due to deaths that occur among mothers at delivery is more difficult to specify. However, by chance it happens that the deaths due to this cause are of the same numerical order as deaths among fathers during the 9 months which precede the birth.* A quite satisfactory correction can thus be computed, although for entirely dif-

* For the details in this regard the reader is referred to my paper, "Orphanhood in relation to demographic factors," *Metron*, 1931, *v. 9*, pp. 37–109, where, however, I follow a much more complicated method, not having yet developed the simple method presented here for children who lose their mother at birth.

ferent reasons, with the same formula (304), giving to $p(a)$, however, values taken from a female life table, and setting $\bar{n}$ equal to the mean age of mothers at the birth of their children.

Absolute Orphans. The case of children who have lost both parents has a particular social importance. If the probability of the loss of the mother and that of the loss of the father were independent, the probability of the loss of both parents would be obtained simply as a product of the two probabilities. Observation alone can tell us if things are this way. As we will see shortly these two constituent probabilities are quite far from being independent.

Applications. 1. Comparison of Theoretical with Observed Numbers: England. To obtain a practical confirmation of the theory, the formulas derived above have been applied to statistical data for England and Wales for the year 1921. For that year we possess complete data on the number of maternal, paternal, and absolute orphans classified by age. The results are presented in Table 15 and Figure 18. It will be seen that in the case of male maternal orphans the accord is quite satisfactory between the computed and observed numbers up to the age of 5.

Above that age the computed numbers are greater than the observed numbers, in part no doubt because of remarriage of some of the widowers, which causes the children not to be considered orphans in the statistics, although represented as such in our formulas.

The number of male paternal orphans reported is, on the contrary, higher than the number computed, particularly at ages 4 to 10, that is to say, for children born between 1911 and 1917. One sees therein one of the fatal effects of the war.

It should be noted as well that the proportion of absolute orphans, computed as the simple product of the proportion of maternal and paternal orphans, does not accord at all with the observations. As a matter of fact, at the age of one year the number of orphans registered is at least ten times greater than the number computed under the hypotheses, evidently very inexact, of independence between the deaths of the father and of the mother.

For older ages the divergence between the computed and observed numbers diminishes more and more.

Table 15. Number of orphans, England and Wales, 1921. Computed versus observed figures

(1)	(2)	(3)	(4)	(5)	(6)	(7)	(8)	(9)	(10)	(11)
		Paternal orphans			Maternal orphans			Absolute orphans		
Age	Total no. of children*	Observed	Computed	Difference	Observed	Computed	Difference	Observed	Computed**	Difference
0–15	10 362 244	786 090	593 501	192 589	316 339	386 746	–70 407	55 245	54 647	598
0–1	785 852	7 541	5 580	1 961	3 382	3 772	- 390	332	236	96
1–2	816 506	13 508	10 615	2 893	5 379	6 940	- 1 561	533	490	43
2–3	544 211	19 780	10 340	9 440	7 048	6 803	245	837	381	456
3–4	529 227	25 000	13 389	11 616	9 088	8 838	250	1 170	582	588
4–5	603 310	36 124	19 185	16 939	11 968	12 730	- 762	1 607	965	642
5–6	647 579	45 577	25 061	20 516	14 765	16 643	- 1 878	2 091	1 295	796
6–7	699 558	57 266	32 040	25 226	18 473	21 197	- 2 724	2 915	1 889	1 026
7–8	716 809	61 932	38 134	23 798	21 519	25 232	- 3 713	3 430	2 509	921
8–9	713 052	63 833	43 496	20 337	23 694	28 593	- 4 899	3 874	3 137	737
9–10	702 427	65 920	48 538	17 382	26 038	31 820	- 5 782	4 543	4 074	469
10–11	707 463	69 423	54 899	14 524	28 394	35 798	- 7 404	5 093	5 023	70
11–12	720 078	73 914	62 287	11 627	31 802	40 396	- 8 594	5 840	6 337	- 97
12–13	732 329	78 954	70 157	8 797	36 004	45 331	- 9 327	6 925	7 689	- 764
13–14	734 055	82 828	77 516	5 312	38 487	49 916	–11 429	7 655	9 322	–1667
14–15	709 788	84 490	82 264	2 226	40 298	52 737	–12 439	8 400	10 718	–2318

* Omitting those for whom the parents' condition was unknown.

** The computed values are corrected for the effect of dependency between the probabilities of death of the two spouses. The base statistics for the computed values are found in the following sources: English Life Table No. 9; England and Wales, Census of 1921, General Tables, Table 33, p. 138; England and Wales, Report of the Registrar General, 1922, p. 138; England and Wales, Census of 1921, Dependency, Orphanhood, Fertility, Table 11, p. 241.

2. Indirect Computation Where Direct Observational Data Are Lacking: United States. In any event, the accord between the results computed according to the theoretical formulas and the observed numbers is sufficiently good to render theoretical numbers useful where direct data on orphans is lacking, as in the case of the United States. In consequence, the theoretical numbers in Table 16 inform us in a certain measure about conditions in the United States at the date of the last census, 1930. In computing the number of absolute orphans we have multiplied the product of the proportions of maternal and paternal orphans by the corresponding factor observed in the population of England and Wales. To be sure, this procedure is not ideal, and it would be very desirable to have data on orphans in other countries to know up to what

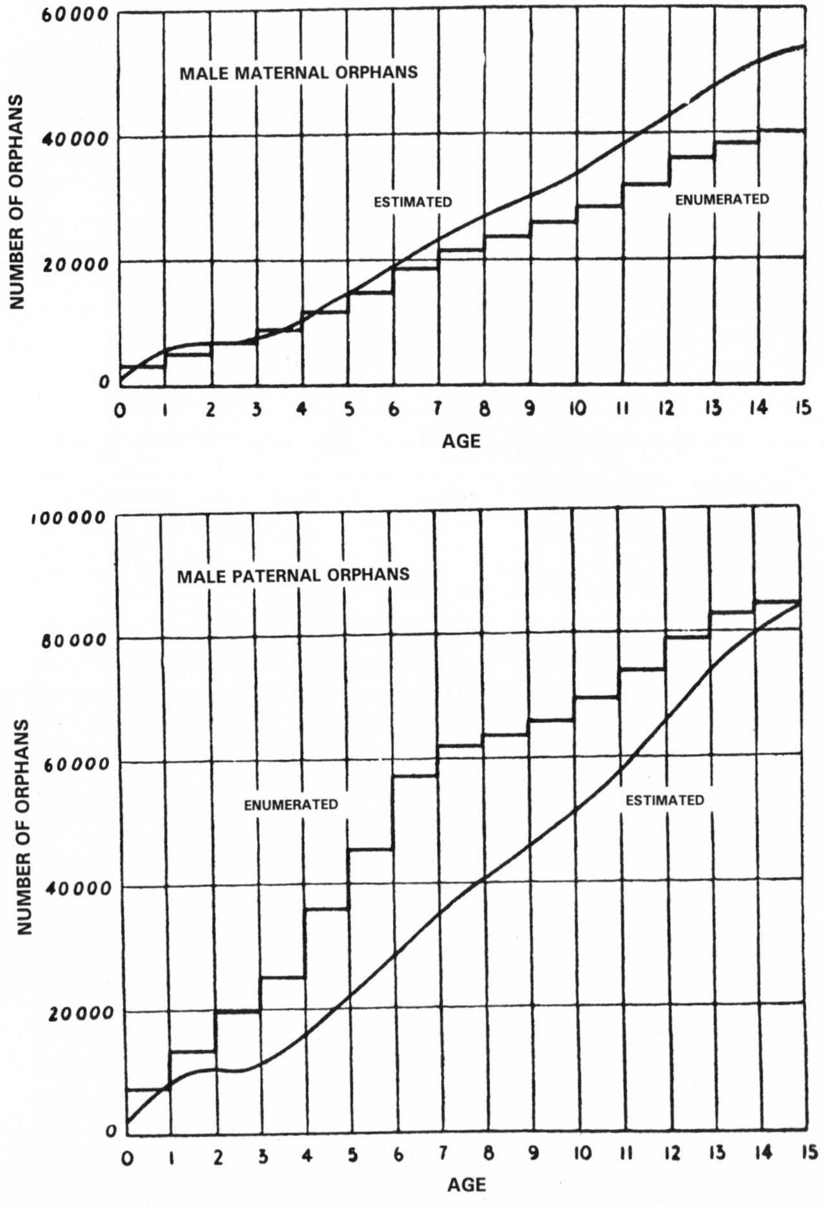

Figure 18. Comparison of observed and computed numbers of orphans, England and Wales, 1921.

point those factors are constant or variable from one case to another. For the moment we must be content to have drawn attention to these relations.

Effect of the Amelioration of Mortality. It is interesting to compute the number of orphans in the same population under the regime of two different rates of mortality, to appreciate the advantage of the diminution of mortality in that respect. The results of such a computation are presented in Table 17. One sees there that if the population of the United States in 1930 had still been under the mortality regime of 1901, there would have been among white children almost 2 million additional orphans.*

Table 16. Estimated number of orphans, United States, white race, 1930

Age	Total number of children	Number of orphans			
		Total number	Father deceased	Mother deceased	Both deceased
Under 16 years	33,448,712	2,718,000	1,531,000	1,041,000	146,000
Under 5 years	9,927,396	266,000	146,000	113,000	7,000
5–9	10,956,144	790,000	443,000	318,000	29,000
10–14	10,546,282	1,332,000	755,000	494,000	83,000

Table 17. Reduction in the number of orphans (under 16 years) due to the decrease in mortality from 1901 to 1930, United States, white race

Condition	Estimated number of orphans in 1930 based on the mortality of		Reduction in the number of orphans
	1901	1930	
Father deceased	2,438,000	1,531,000	907,000
Mother deceased	1,741,000	1,041,000	700,000
Both deceased	455,000	146,000	309,000
Total	4,634,000	2,718,000	1,916,000

* See A. J. Lotka, "Orphanhood in relation to demographic factors," *Metron*, 1931, *v. 9*, p. 37; also M. Spiegelman, "The broken family: Widowhood and orphanhood," *Annals of the American Academy of Political and Social Sciences*, 1936, *v. 188*, p. 126.

Composition of the Family by Number of Children. The same method which has furnished us the values of $m_1(a)$, the number of first births per woman of age a, presents as well the data for computing $m_2(a)$, $m_3(a)$, ..., $m_n(a)$, — the number of second, third, ..., nth births per woman of age a. Just as we followed the life of a cohort of N women to determine the proportion of wives and mothers, we can look for the proportion among them who will have 2, 3, ... children under the regime of a given fertility and mortality table. We obtain by this means the classification of families according to the number of children born alive. To compute, for example, the number of women in the cohort who are mothers of 3 children, neither more nor less, we form the difference

$$N\int_0^{\omega} p(a)\,\{m_3(a)-m_4(a)\}\,da = N\,p_3 \qquad (309)$$

Application. In this example we have utilized statistics for the white population of the United States for two years, namely 1920 and 1930. The results represented in Figure 19, using a logarithmic scale for the ordinates, make evident a remarkable property, and one that is very useful in the discussion of the problem of the extinction of families. The points giving the relative frequency of families of 1, 2, ..., n children are distributed nearly in a straight line;* In fact, designating by p_n the relative frequency of families of n children per mother, we find by the method of least squares the very reasonable approximation

$$p_n = 0.7358^{(n-1)}\,p_1 \qquad (n \neq 0) \qquad (310)$$

In general the frequency of families without children forms an exception and does not enter into the linear logarithmic schema. Let us designate by p_0, p_1, p_2, ..., p_ω the probability of having 0, 1, 2, ..., ω children in the family. These probabilities must satisfy the condition

---

* This property had been signaled for male children by A. J. Lotka, *Journal of the Washington Academy of Sciences*, 1931, pp. 377–453. An analogous observation for children of both sexes is found in E. B. Wilson, *Journal of the American Statistical Association*, 1935, *Supplement*, p. 577. The two authors have remarked that the linearity ceases to be exact beyond families of 11 or 12 children.

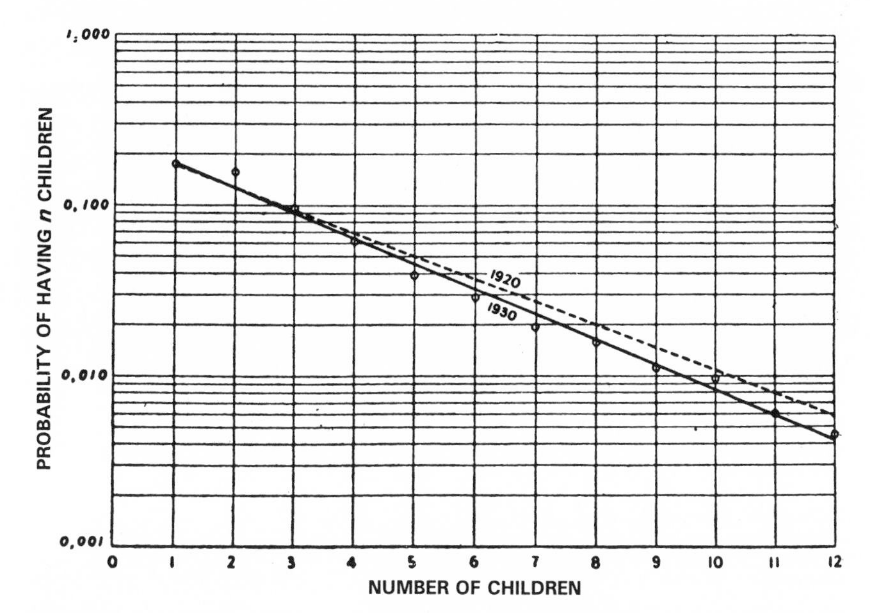

Figure 19. Probability that a newborn will have n children in the course of his life. Based on statistics for the United States in 1920 and 1930. The small circles indicate the actual statistics for 1930; the lines are constructed by the method of least squares.

$$p_0 + p_1 + p_2 + \cdots + p_\omega = 1 \tag{311}$$

which, in general, would not be possible if all of the terms in p formed a geometric series. On the other hand, after having computed by the method of least squares a straight line representing the collection of points p_1, p_2, ..., p_ω, the corresponding value of p_0 can always be computed by the formula

$$p_0 = 1 - (p_1 + p_2 + \cdots + p_\omega) \tag{312}$$

CHAPTER 8

Extinction of a Line of Descent

The reasoning by which we have arrived at the notion and measure of the intrinsic rate of increase of the population is based on the general fertility function $m(a)$. In consequence, our conclusions represent the mean, the collective result as it presents itself in a population taken as a whole. The functions of fertility by the order of the child $m_n(a)$ that we have just introduced permit us to push our analysis further. The average rate of increase r does not apply equally to any element of the population taken randomly, for example to the descendants in a direct line (carrying the same family name) of some individual.

Let us study a special problem: what is the probability that the number of (male) descendants in a direct line to the sth generation for some (male) individual will be equal to ν? What, in particular, is the probability that this number will be zero; that is, that the family name will be extinguished after the sth generation? That this problem, in one form or another, is of considerable interest is evident by the number of authors who have been occupied with it. The list includes the names of Watson (1889), cited by Galton (1889),[*] Erlang, cited by Steffensen (1930),[†] Lotka (1931),[‡] J. B. S. Haldane (1932),[§] and R. A. Fisher (1929),[**] the latter two authors being occupied with the application of that problem to certain questions of evolution under the regime of Mendelian heredity.

The application of theoretical formulas of the type of those of Watson and Steffensen to concrete statistical data was initially made by Lotka (1931) and independently by Steffensen (1931). This practical

---

[*] H. W. Watson cited by F. Galton, *Natural Inheritance*, 1889, p. 242.

[†] J. F. Steffensen, *Mathematisk Tidskrift*, 1930, p. 19.

[‡] A. J. Lotka, *Journal of the Washington Academy of Sciences*, 1931, pp. 377, 453.

[§] J. B. S. Haldane, *The Causes of Evolution*, 1932, p. 198.

[**] R. A. Fisher, *The Genetical Theory of Natural Selection*, 1929, p. 273.

application requires certain special expansions, as we will see.

We will study the male descendants of a male individual from the moment of his birth.

Let π_0 be the probability considered at the moment of his birth* that an individual dies without male progeny;†

π_1 be the probability that he has had a single son before his death;

π_2 be the probability that he has had two sons (neither more nor less) before his death;

π_ν be the probability that he has had ν sons (neither more nor less) before his death.

Then, evidently

$$\pi_0 + \pi_1 + \pi_2 + \cdots + \pi_\omega = 1 \tag{313}$$

ω being the largest number of sons for which the probability π_ω is non-zero.

We will treat only the case where the π are invariable for the individual parent and all of his descendants.

We form an auxiliary function

$$f(x) = \pi_0 + \pi_1 x + \pi_2 x^2 + \cdots \tag{314}$$

having certain remarkable properties that will be useful to us. We remark in particular that

1st $f(1) = 1$ (315)

2nd $f(0) = \pi_0$ = probability that the individual dies without any male descendant (316)

---

* J. F. Steffensen, "Deux problèmes du calcul de probabilité," *Conferences held at the Poincaré Institute*, December 1931; see also E. B. Wilson, *Journal of the American Statistical Association*, 1935, *Supplement*, p. 577.; T. H. Rawles, *Human Biology*, 1936, p. 126.

P. L. Fegiz, "I cognomi de San Gimignano," *Metron*, 1925, *v. 5*, p. 115.

† Stillbirths being excluded.

3rd $\left(\frac{\partial f}{\partial x}\right)_1 = f'(1) = \pi_1 + 2\pi_2 + 3\pi_3 + \cdots$ (317)

= mean number of sons of the individuals in question

4th The coefficient of x^s in the series $f(x)$ indicates the probability for an individual (considered at the moment of his birth) of having s sons in the course of his life.

5th The coefficient of x^s in the series $[f(x)]^r$ gives the probability that r individuals (counted at the moment of their birth) will together have s sons in the course of their lives (the reader can assure himself of the validity of this proposition by computing this probability according to the usual rules and comparing the result with the coefficient in question).

6th The probability for an individual of having r sons and s grandsons is expressed by the product of the probability that he has r sons and the probability that these r sons have in turn s sons; this compound probability is given by the coefficient of x^s in the product

$$\pi_r [f(x)]^r \tag{318}$$

7th The probability for an individual of having s grandsons, whatever the number of his sons, is evidently given by the coefficient of x^s in the sum of all the terms of the form $\pi_r\ [f(x)]^r$, that is, in

$$\pi_0 + \pi_1 f(x) + \pi_2 f^2(x) + \cdots + \pi_\omega f^\omega(x) = f_2(x) \tag{319}$$

This function

$$f\{f(x)\} = f_2(x)$$

and in general

$$f\{f[f \cdots]\} = f_n(x) \tag{320}$$

is called the iterated function of order n of $f(x) = f_1(x)$.

8th By the same reasoning as in 7 the reader can assure himself that the probability for an individual of having ν sons in the sth generation is given by the coefficient of x^{ν} in the sth iteration $f_s(x)$ of $f(x)$.

It still remains to determine what the function $f_s(x)$ becomes when s tends to ∞, and particularly what then happens to $f_s(0)$, that is, what will be the probability that the line of descent is extinguished.

To resolve this question we must examine certain properties of the iterations of $f(x)$.

Properties of the Iterated Functions $f_s(x)$. First, the coefficients π being all positive quantities, and their sum Σ π being equal to 1, it follows that $f(x)$ as well as all of its iterations increase continuously from π_0 to 1 when x increases from 0 to 1.

Second, if there exists a value of x, let us say $\xi < 1$ such that

$$\xi = f(\xi) \tag{321}$$

then, evidently, we will also have

$$\xi = f_1(\xi) = f_2(\xi) = \cdots = f_{\omega}(\xi) \tag{322}$$

These two properties are represented in Figure 20, which gives both a graphic method of construction for the successive iterations of $f(x)$, and the demonstration of a theorem which permits computation of the limiting value of $f_n(0)$ for n tending to ∞, that is to say, the probability of extinction of a line of descent.

Theorem on the Limiting Value of a Function Iterated an Infinite Number of Times. The limiting value of $f_n(x)$ for n tending to ∞ is nothing else than the value ξ of x which satisfies the condition

$$\xi = f_1(\xi)$$

provided that such a value ≤ 1 exists.

The meaning of the theorem is clear: it suffices to regard Figure 20, where the first function $f(x) = f_1(x)$ has been traced directly; to con-

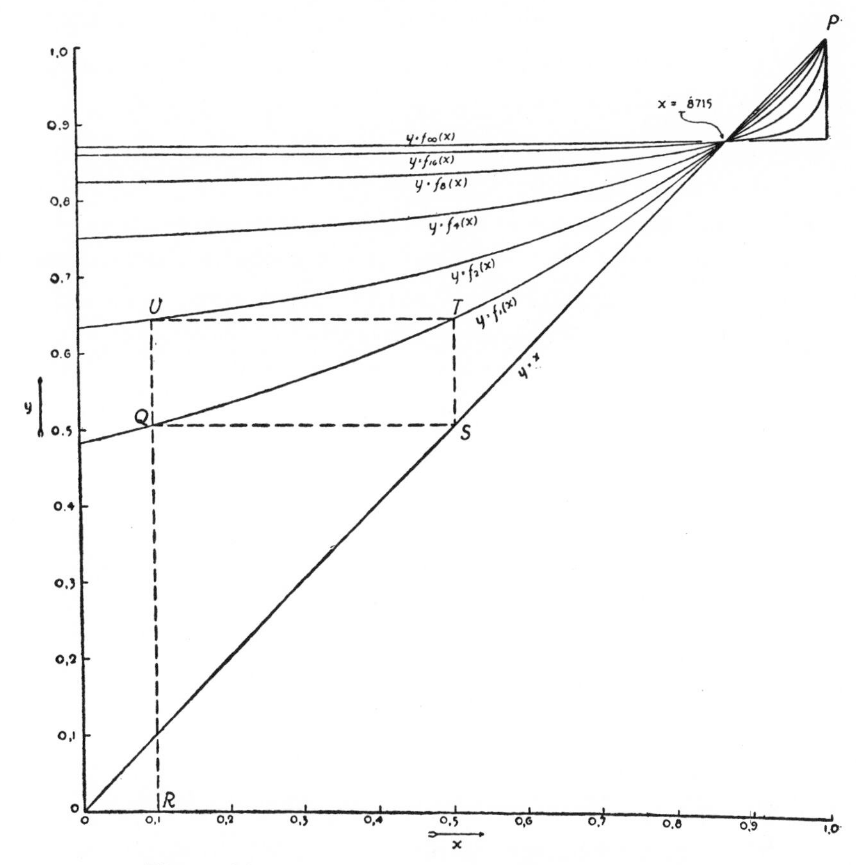

Figure 20. Graphic iteration of the function $y = f_n(x)$.

struct it's iteration $f_2(x)$ we have drawn from some point Q on the curve $y_1 = f_1(x)$ a line parallel to the x axis and cutting the diagonal OP ($y = x$) at the point S. The ordinate of the point S cuts the curve $y_1 = f_1(x)$ at the point T, and the horizontal line passing through T cuts the ordinate of the point Q at the point U, which is evidently situated on the curve $y_2 = f_2(x)$. By proceeding in this fashion we can construct any number of points of the curve $y_2 = f_2(x)$.

By operating in the same manner using the curve $y_2 = f_2(x)$, we obtain the curve $y_3 = f_3(x)$, and so forth. We have already remarked that the function $f_1(x)$ and all its iterations are increasing functions of x. Thus, if there exists a value ξ (< 1) of x, such that $\xi = f(\xi)$, the curve $f_1(x)$ and all its iterations pass through that point, located at the intersection of the diagonal with the curve $f_1(x)$. Moreover, all of these curves pass through the point $x = y = 1$.

Due to the way the curves are constructed, the parts of the curves situated to the left of the point $\xi = f_1(\xi)$ are located one above the other, flattening and approaching more and more closely to a horizontal line, which represents the corresponding part of the limiting curve $y = f_\infty(x)$. It follows that the curves that represent $f_1(0)$, $f_2(0)$, ..., $f_\infty(0)$ are arranged one above the other, tending toward the limit

$$f_\infty(0) = f_\infty(\xi) = f_1(\xi) = \xi \tag{323}$$

Therefore, if we want to know the probability that the direct (male) line of descent of an individual is extinguished, it is necessary to look for the root < 1 (if it exists) of the equation

$$x = f(x)$$

In consequence, we must first of all examine more closely the function $f(x) = \Sigma\, \pi_i x_i$.

Let P_n be the probability at his birth for a male individual to have in the course of his life at least n children. We know that under these conditions

$$P_n = \int_0^\infty p(a)\, m_n(a)\, da \tag{324}$$

$$P_0 = 1 \tag{325}$$

the functions $p(a)$, $m_n(a)$ being with respect to the male sex (fathers).

The probability that there will be n children, neither more nor less, is evidently

$$p_n = P_n - P_{n+1} \tag{326}$$

We will look for the corresponding probability that there will be ν children *of the male sex*.

Let α be the proportion of male births and $\beta = 1 - \alpha$ the proportion of female births among all births. The probability that among n children there will be ν sons is then given* by the coefficient of α^ν in the series expansion of $(\alpha + \beta)^n$.

What then will be the probability for a newborn male of having in the course of his life ν sons? The answer is given by Table 18. In the last line we have inscribed the probability for a newborn of having 0, 1, 2, ... children, under the hypothesis, justified by the observations already cited, that these probabilities† form a geometric series, with a constant ratio γ. In each column of the table the probability of having n children, which is found at the bottom of the column, has been decomposed according to a well known theorem in probability that among those n children there are 0, 1, 2, ... sons. The eighth column gives the values of the probability π_ν of having ν sons independent of the total number of children. It will be seen immediately that if the probabilities $p_1, p_2, \ldots, p_\omega$ of having 1, 2, ..., ω children form a geometric series with a constant ratio γ, the corresponding probabilities $\pi_1, \pi_2, \ldots, \pi_\omega$ of having 1, 2, ..., ω sons, also form a geometric series whose constant ratio is equal to

$$\Gamma = \frac{\alpha\gamma}{1 - \beta\gamma}$$

In practice α and β do not diverge much from the value $1/2$, and the constant ratio of the series $\pi_1, \pi_2, \ldots$ is thus $\gamma/(2 - \gamma)$.

Mean Number of Sons. The mean number of sons eventually born per newborn male infant (in other words, the mathematical expectation

---

* By a well known theorem in the theory of probabilities. We have supposed here that the probability of births of each sex observed among total births also applies in each family. This hypothesis does not accord exactly with observations, but the error that it introduces is unimportant. A very detailed statistical study on this subject will be found in A. Geissler, *Zeitschrift des K. Sachsischen Statistischen Bureaus*, 1889, *books I–II*. See also D. Cecil Rife and L. H. Snyder, *Human Biology*, 1937, *v. 9*, p. 98.

† Except that of 0 children (see pp. 178–179).

of the eventual number of sons for a newborn boy) is evidently obtained upon multiplying by 1, 2, 3, ... the second, third, fourth, ... term in the eighth column of Table 18, and forming the sum. We obtain in this way

$$m = \frac{\alpha p_1}{(1-\gamma)^2} \tag{327}$$

a result that is evidently in accord with the mean number of children found by forming the sum

$$p_1 + 2\gamma p_1 + 3\gamma^2 p_1 + \cdots = \frac{p_1}{(1-\gamma)^2} \tag{328}$$

Probability of the Extinction of a Line of Descent. Following the principles expounded earlier, this probability is given by the root < 1 of the equation

$$x = \pi_0 + \pi_1 x + \Gamma \pi_1 x^2 + \Gamma^2 \pi_1 x^3 + \cdots \tag{329}$$

$$= \pi_0 + \frac{\pi_1 x}{1 - \Gamma x} \tag{330}$$

We already know that the sum of the probabilities of having 0, 1, 2, 3, ... sons is equal to 1; that is

$$1 = \pi_0 + \frac{\pi_1}{1-\Gamma}. \tag{331}$$

On eliminating π_1 from (330) and (331) we obtain

$$x(1-\pi_0)(1-\Gamma) = (x - \pi_0)(1 - \Gamma x) \tag{332}$$

$$x^2 - x\left(\frac{\pi_0}{\Gamma} + 1\right) + \frac{\pi_0}{\Gamma} = 0 \tag{333}$$

$$(x-1)\left(x-\frac{\pi_0}{\Gamma}\right)=0 \tag{334}$$

The root $x = 1$ does not interest us, we know that it must exist.* That which we are seeking is the root

$$x=\frac{\pi_0}{\Gamma}=\xi \tag{335}$$

provided that it is < 1, that is, provided that $\Gamma > \pi_0$.

Application. We apply these results to statistics for the United States population (white race) in 1920, for which an earlier study† contained the fundamental information. This information and the results obtained by the application of the schema of Table 18 are displayed together in Table 19. The numbers in the table correspond, line by line and column by column, to the quantities represented by the symbols of Table 18. The same information is found in graphic form in Figure 21, where

---

* *Translators' note*: The restriction of the solution to the root $x < 1$ (or to $\pi_0 \leq x < 1$, since the probability of extinction cannot be less than the probability that the individual has no sons) corresponds to the restriction $\Sigma i\pi_i = R_0' > 1$, and to the special case $R_0' = \pi_1 = 1$, where the prime designates the male net reproduction rate, used by Lotka in connection with expression (341). For all other cases $R_0' \leq 1$, the probability of extinction is unity, the solution being the root $x = 1$. Watson had assumed that the solution $x = 1$ held generally, not recognizing the significance of a possible second root in the interval $(\pi_0, 1)$. Such a root exists if $R_0' > 1$ or $[\pi_2 + 2\pi_3 + \cdots + (\omega-1)\pi_\omega] > \pi_0$ (that is, if the number of births of order 2 and higher exceeds the probability of no births), or if all individuals have exactly one son. The correct solution was first given by Steffensen.

† A. J. Lotka, "The extinction of families," *Journal of the Washington Academy of Sciences*, 1931, *v. 21*, pp. 377, 453. This study is developed along similar lines, which, however, have been developed a little further here. The new results differ a little from those in the earlier article: first, in the article the computation is based directly on the distribution of the probabilities of having *n* children, without attributing to them a (logarithmic) rectilinear pattern; such a distribution had been applied only to the probabilities of having ν sons; second, the article did not contain the examination of the relation between these two distributions. The results of this examination are presented here for the first time; third, in the article we set $\alpha = \beta = 1/2$ while in the present text more exact values are used, namely, $\alpha = 0.515$; $\beta = 0.485$.

Table 18. Probability that a male (considered at the moment of his birth) will have n children comprising ν sons, given that the proportion of male births is α and that of female births is $\beta = 1 - \alpha$

Sons, ν	Children, n						$\sum_0^\infty$	Value of constants (see p. 192)
	0	1	2	3	4	5		
0	p_0	βp_1	$\beta^2\gamma p_1$	$\beta^3\gamma^2 p_1$	$\beta^4\gamma^3 p_1$	$\beta^5\gamma^4 p_1$	$p_0 + \frac{\beta p_1}{1-\beta\gamma} = \pi_0$	$\pi_0 = 0.4825$
1		αp_1	$2\alpha\beta\gamma p_1$	$3\alpha\beta^2\gamma^2 p_1$	$4\alpha\beta^3\gamma^3 p_1$	$5\alpha\beta^4\gamma^4 p_1$	$\frac{\alpha p_1}{(1-\beta\gamma)^2} = \pi_1$	$\pi_1 = 0.2126$
2			$\alpha^2\gamma p_1$	$3\alpha^2\beta\gamma^2 p_1$	$6\alpha^2\beta^2\gamma^3 p_1$	$10\alpha^2\beta^3\gamma^4 p_1$	$\frac{\alpha^2\gamma p_1}{(1-\beta\gamma)^3} = \pi_1\Gamma$	$\alpha = 0.515$
3				$\alpha^3\gamma^2 p_1$	$4\alpha^3\beta\gamma^3 p_1$	$10\alpha^3\beta^2\gamma^4 p_1$	$\frac{\alpha^3\gamma^2 p_1}{(1-\beta\gamma)^4} = \pi_1\Gamma^2$	$\beta = 0.485$
4					$\alpha^4\gamma^3 p_1$	$5\alpha^4\beta\gamma^4 p_1$	$\frac{\alpha^4\gamma^3 p_1}{(1-\beta\gamma)^5} = \pi_1\Gamma^3$	$\gamma = 0.7358$
5						$\alpha^5\gamma^4 p_1$	$\frac{\alpha^5\gamma^4 p_1}{(1-\beta\gamma)^6} = \pi_1\Gamma^4$	$\Gamma = \frac{\alpha\gamma}{1-\beta\gamma}$
6								$= 0.5893$
7								
8								$p_0 = 0.3686$ $p_1 = 0.1707$
9								$m = 2.446$ $M = 1.2600$
∞							$p_0 + \frac{p_1}{1-\gamma} = \pi_0 + \frac{\pi_1}{1-\Gamma}$	$\xi = \frac{\pi_0}{\Gamma} = 0.819$

the points marked by small circles represent on a logarithmic scale the probability for a newborn male of having in the course of his life 1, 2, 3, ..., n children. The straight line passing through these points is computed by the method of least squares, leaving aside, however, the points corresponding to children of order 13 and higher, which diverge noticeably from the straight line. (Families that large being quite rare, the error introduced is not serious.) Finally, the dashed line represents, for the abscissas 1, 2, 3, ..., ν, the probabilities according to the formulas of Table 18 for a newborn male of having 1, 2, 3, ..., ν sons in the course of his life. The points corresponding to no children and no sons were computed separately to verify the evident condition that the total sum of the probabilities is equal to 1.

Figure 21 shows in particular that the probabilities of having 0 or 1 son are greater than those of having 0 or 1 child; and that, on the contrary, the probabilities of having 3 or more sons are smaller than those of having the same number of children. The probability of having 2 sons is almost equal to that of having 2 children. One is naturally curious to understand the reason for this. The probability of having 2 sons is

$$\pi_2 = \frac{\pi_1 \alpha \gamma}{(1-\beta\gamma)} = \frac{\alpha^2 \gamma p_1}{(1-\beta\gamma)^3} \tag{336}$$

Now, α and β have approximately the value $1/2$, and γ, according to the data of our example, has the value 0.736, let us say $3/4$. The ratio that we seek is thus, approximately

$$\frac{\pi_2}{p_2} = \frac{\alpha^2 \gamma p_1}{(1-\beta\gamma)^3 \gamma p_1} = \frac{\alpha^2}{(1-\beta\gamma)^3} = \frac{1/4}{(5/8)^3} \tag{337}$$

$$= \frac{512}{500} = 1.024 \tag{338}$$

Its more exact value is

$$\frac{(0.515)^2}{(1-0.485 \times 0.736)^3} = \frac{0.2652}{0.2660} = 0.9971 \tag{339}$$

Table 19. Probability that a newborn male will have, in the course of his life, *n* children comprising ν sons. Computations based on data for the United States, white race, 1920

Sons, ν	Children, *n*											
	0	1	2	3	4	5	6	7	8	9	10	$\sum_{n=0}^{n=\infty} p_{n,\nu}$
0	0.3686	0.0828	0.0295	0.0105	0.0038	0.0013	0.0005	0.0000	0.0000	0.0000	....	0.4825
1		0.0870	0.0628	0.0336	0.0160	0.0071	0.0030	0.0013	0.0005	0.0002	....	0.2126
2			0.0333	0.0357	0.0254	0.0152	0.0081	0.0041	0.0019	0.0009	....	0.1252
3				0.0126	0.0180	0.0161	0.0115	0.0072	0.0041	0.0022	....	0.0738
4					0.0048	0.0086	0.0091	0.0076	0.0055	0.0035	....	0.0435
5						0.0018	0.0039	0.0049	0.0046	0.0037	....	0.0257
6							0.0007	0.0017	0.0025	0.0026	....	0.0151
7								0.0003	0.0007	0.0012	....	0.0088
8									0.0001	0.0003	....	0.0052
9										0.0000	....	0.0031
10											....	0.0018
$\sum_{\nu=0}^{\nu=n} p_{n,\nu}$	0.3686	0.1707	0.1256	0.0924	0.0680	0.0501	0.0368	0.0271	0.0199	0.0147	....	1.0000

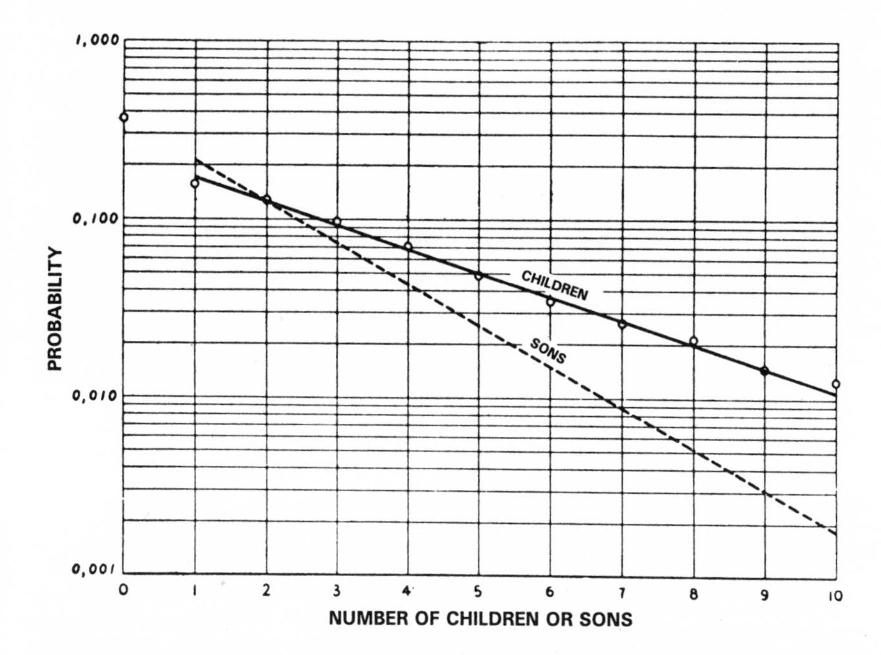

Figure 21. Probability that a newborn male will have a given number of children or sons in the course of his life. Based on data for the United States, 1920.

In rounded numbers the ratio is thus 1, which explains why the probability of having 2 sons is practically the same as that of having 2 children,based on the conditions of our example taken from statistics for the American population.

Mean Number of Sons: Numerical Example. After formula (327) the mean number of sons is found to be

$$m = 1.260 \tag{340}$$

which is equal to the ratio of total male births in two successive generations. We recall that, based on the same statistics, the ratio of female births was $R_0 = 1.166$. The difference is due in part to the fact that the mean interval between father and son is longer than that between mother and daughter; the figures are 32.76 and 28.33 years, respectively. We can compute a value R_0' for males corresponding to R_0 for females, based on the relation

$$\frac{\ln R_0'}{T'} = \rho \tag{341}$$

We have found for ρ the value 0.0054. It follows that

$$\ln R_0' = 0.0054 \times 32.76 = 0.17$$
$$R_0' = 1.194 \tag{342}$$

a value that is also sensibly inferior to the value 1.260.

The fact is that a certain disparity between these numbers must be expected. Not only is the mean number m computed by a fairly indirect process, including among other things an adjustment by the method of least squares, but in addition we have summed our series to infinity, while families above 30 or 40 children, for example, certainly do not exist. If the computation is stopped at families of 20 children, including 10 sons, the value of m is found to reduce to 1.232. Stopping at families of 15 children, including 10 sons, the value of m becomes 1.198. In all of these computations we have supposed that the frequency of n children per family follows exactly the law of a geometric series, although we know that beyond the 12th child the frequencies do not reach the values given by this series.

Finally, although ultimately the male and female populations must have equal rates of increase (without that, one would finally infinitely surpass the other), during a fixed period it can certainly happen, and in general there will be a certain discord between the two rates of increase. The approximate concordance between $m = 1.260$ computed for males, $R_0 = 1.166$ computed for females, and $R_0' = 1.194$ computed for males is thus as close as circumstances permit us to expect.

Probability that the *S*th Generation Comprises ν Sons. Following the principles already enunciated the probability in question can be computed by expanding the *s*th iteration of $f(x)$, and choosing the coefficient of x^ν in the series that results. The successive iterations will give increasingly accurate results, but thanks to a formula by Steffensen* the desired coefficient can be obtained by a direct route. This formula, slightly modified, has the form

$$\pi_{s,\nu} = m^{s}\left(\frac{1-\xi}{m^{s}-\xi}\right)^{2}\left(\frac{m^{s}-1}{m^{s}-\xi}\right)^{\nu-1} \tag{343}$$

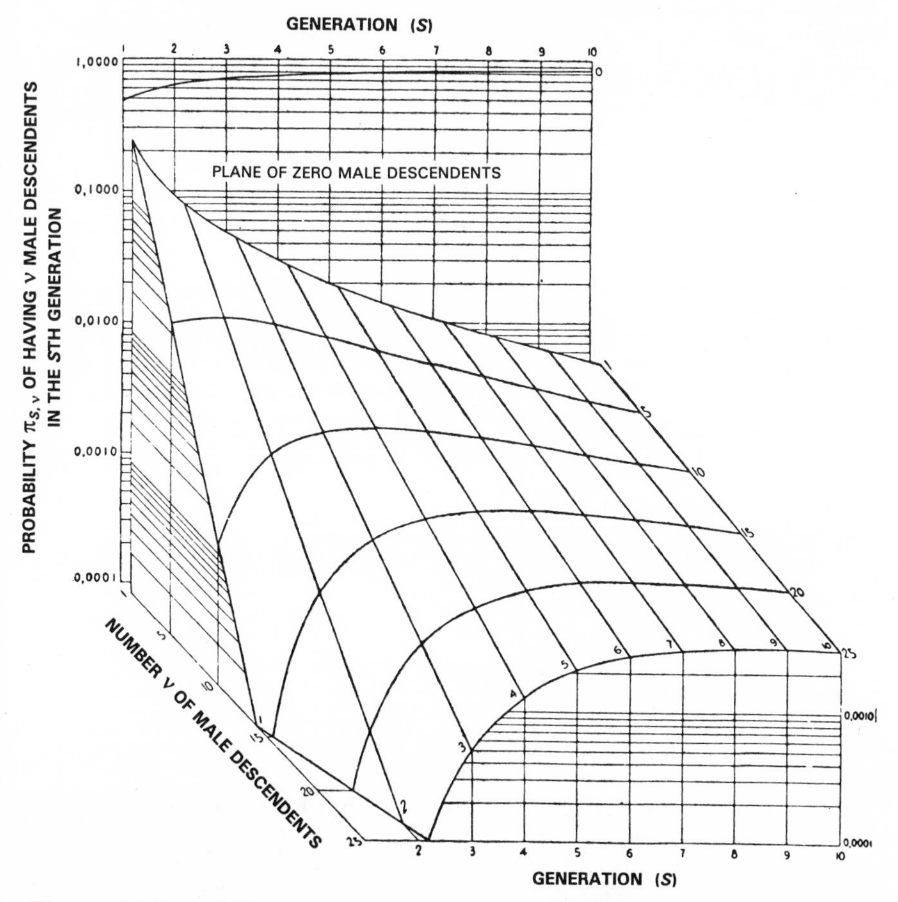

Figure 22. Probability $\pi_{s,\nu}$ that a newborn male will have ν male descendants in the *s*th generation. Logarithmic scale.

---

* *Loc. cit.*, p. 342. The modification cited here applies to the case where the probabilities of having a given number (except none) of sons form a geometric series.

It links the probability $\pi_{s,\nu}$ with the probability ξ of the extinction of the line, and the mean number m of sons in the first generation.

A good idea of the character of this relation can be obtained by examining the graph (Figure 22), which gives a representation of it in three dimensions.†

One additional result, cited by Steffensen, merits brief mention. In the special case where the entire series of probabilities $\pi_0, \pi_1, ..., \pi_\omega$ forms a geometric series,* there is a very simple relation between the probability of extinction ξ and the mean number of sons in the first generation m,†

$$\xi = \frac{1}{m} \tag{344}$$

† This graph is taken from the author's article, "The extinction of families," already cited, in which the computations are based on an adjustment by the method of least squares of the distribution of descendants in the first generation according to number of sons, while here the adjustment is applied to the distribution by number of children. The difference in the results is inconsiderable and is unimportant from a topological perspective.

* That is to say, without excepting π_0.

† J. F. Steffensen, loc. cit., p. 343. [*Translators' note*: In Lotka's expression (344), $\xi = 1$ for $m \leq 1$.]

CHAPTER 9

Conclusion

Within the limits imposed on monographs in this series, the author has contented himself with a summary of recent progress in demographic analysis. The classical sources, which in any event are cited in all collective works in quantitative demography, are left aside to avoid repetition of what is sufficiently well expounded elsewhere. To complete the plan of the present monograph it remains only to indicate a device to which the author has alluded in the first pages. We had remarked there that graphic representations, while rendering good service, are insufficient for expressing the complicated relations of demographic analysis, not only because of the large number of variables that enter into play, but even more because of the form of these relations. To effectively represent probabilistic relations it is not immobile graphics inscribed on paper that we need. We should rather construct mobile models which by their operation produce a collection of characteristics having a distribution of frequencies resembling the characteristics of the population to which they apply. Our ideal would be a collection of urns giving, by a draft effected in accord with prescribed rules, a true sampling of the characteristics of the population.* That ideal is unrealizable in every detail, but it can be approached to a degree, depending on the one hand on our statistical resources, and on the other on the effort we are ready to give to the job.

---

* In this connection we cite Emile Borel: "Le problème général de la statistique est le suivant: déterminer un système de tirages effectués dans les urns de composition fixe, de telle manière que le résultat d'une série de tirages, interprétés à l'aide de coefficients fixes convenablement choisis puissent avec une très grande vraisemblance conduire à un tableau identique au tableau des observations." [The general problem of statistics is the following: to determine a system of drawings effected in urns of fixed composition, in such a manner that the result of a series of drawings, interpreted with the aid of suitably chosen fixed coefficients, can with a very high probability conduct us to a table identical to a table of observations.] (*Le Hasard*, 1928, p. 154.)

By way of example we will imagine such a collection of urns, sufficiently complete to represent with a certain approximation the most fundamental properties and relations of the population, but naturally without taking into account certain details for which we presently lack numerical data.

We remark at the outset that urns *per se* would be inconvenient, in very badly cluttering our bookshelves. We will replace them by a finer equivalent.

We will have, first, a disk, divided by radial lines into sectors proportional to the numbers of the d_x column of a life table for the male sex.

Let us suppose that we are in the presence of a certain number of newborn boys. Let us spin the disk, and, stopping it by a convenient mechanism that functions in an impartial manner, let us note the age corresponding to the sector that is found to rest against a fixed pointer. For example, let that age be 52 years. We say then that one of our boys has lived up to the age of 52. We can repeat this operation indefinitely and construct an entire cohort of individuals, among whom the decedents will be distributed according to the life table utilized in the disk's construction. Actually, this cohort can be drawn directly from the life table. But, let us continue. We choose one of these individuals, for example the one who was dead at age 52. We now take a second disk which we spin and which we stop by chance like the first. This time the division in sectors is such that the point at which it stops indicates to us the age at which this individual married, let us say 25 years. (If by chance the disk stops at age 54 it can be said that this individual has died without having been married.) A third disk will indicate in an analogous manner the age of his spouse, let us say 22 years, at the marriage. We take a disk similar to the first but constructed on the basis of the female life table. This disk being spun and stopping by chance will tell us at what age the woman will die. If by chance the disk stops at an age less than 22 we ignore the outcome and repeat the process until we obtain an age greater than 22. Let us suppose we have obtained the age 40. We pass to a fifth disk. One year after the marriage, the woman being age 23, this disk indicates to us on its first spin if she gives birth to a child. A second spin gives us the same information with respect to the second year of marriage, when the woman is 24, and so forth, up to age 40, the age at death of this woman. The sex of the child is determined each time by manipulation of a disk having two sectors in the proportions 0.515 and 0.485.

By this means a population can be constructed having characteristics corresponding as exactly to an actual population as there are sufficient disks in front of us corresponding faithfully to the demographic characteristics of this population. The series of disks taken as an example in what has preceded represents the minimum necessary for a model to some degree realistic for a human population. The author has constructed such a series of disks of about 30 centimeters in diameter. Unfortunately, we can not reproduce this model of the population of the United States in the format of this book.

* * *

And now, to complete this presentation of recent results in demographic analysis, we recall that this analysis, bearing on phenomena of a population of a single species, is only a special chapter in the general analytical theory of biological associations which was our point of departure, and to which we must return to complete the study that fundamentally interests us.

Appendix

Change of the variable in a series expansion by means of the successive derivatives of a given function (see p. 92). Let

$$\Phi(t) = c_0\,\varphi(t) + c_1\,\varphi'(t) + \frac{1}{2!}c_2\,\varphi''(t) + \cdots \tag{345}$$

We seek an expansion

$$\Phi(t) = c_0'\varphi(0) + c_1'\varphi'(0) + \frac{1}{2!}c_2'\varphi''(0) + \cdots \tag{346}$$

$$= c_0'\varphi(t+\tau) + c_1'\varphi'(t+\tau) + \frac{1}{2!}c_2'\varphi''(t+\tau) + \cdots$$

$$= c_0'\left\{\varphi(t) + \tau\varphi'(t) + \frac{1}{2}\tau^2\varphi''(t) + \cdots\right\} +$$

$$c_1'\left\{\qquad \varphi'(t) + \quad \tau\varphi''(t) + \cdots\right\} +$$

$$\frac{1}{2!}c_2'\left\{\qquad\qquad \varphi''(t) + \cdots\right\} + \cdots$$

$$\begin{aligned} &= c_0'\varphi(t) + (c_0'\tau + c_1')\varphi'(t) + \\ &\frac{1}{2!}(c_0'\tau^2 + 2c_1'\tau + c_2')\varphi''(t) + \cdots \end{aligned} \tag{347}$$

Comparing the terms of series (345) with those of series (347) we recognize without difficulty the law of formation of the coefficients c' in the expansion of $\Phi(t)$ in terms of $\varphi(t)$ and its derivatives; these are,

$$c_0 = c_0'$$
$$c_1 = c_0'\tau + c_1'$$
$$c_2 = c_0'\tau^2 + 2c_1'\tau + c_2'$$
$$c_3 = c_0'\tau^3 + 3c_1'\tau^2 + 3c_2'\tau + c_3'$$

etc.

Bibliography

Note: The bibliography is not intended to be complete or detailed, but is rather to orient the reader: the six subtitles of the table correspond to fundamental divisions of our subject. Several of these publications contain bibliographic notes as well, and are useful complements to our necessarily limited choice. It goes without saying that it is not always possible to trace the boundaries between the subdivisions of a classification well.

1. General Demographic Analysis

Bonz, F. and Hilburg, F. Die voraussichtliche Bevölkerungsentwicklung in Deutschland. *Zeitschrift für angewandte Mathematik und Mechanik*, 1931, *v. 11*, p. 237.

von Bortkiewicz, L. Mittlere Lebensdauer. *Staatswissenschaftliche Studien*, 1893, *v. 4*.

von Bortkiewicz, L. Die Sterbeziffer und der Frauenüberschuss in der stationären und in der progressiven Bevölkerung. *Bulletin de l'Institut International de Statistique*, 1911, *v. 19*, p. 63.

Connor, L. R. Fertility of marriage and population growth. *Journal of the Royal Statistical Society*, 1926, *v. 89*, p. 553.

Friedli, W. Bevölkerungsstatistische Grundlagen zur Alters- und Hinterlassenenversicherung in der Schweiz. *Bundesamt für Sozialversicherung*, Bern, 1928, p. 32.

Gumbel, E. J. La durée extrême de la vie humaine. *Actualités Scientifiques et Industrielles*, No. 250, Paris: Hermann, 1937.

Haldane, J. B. S. A mathematical theory of natural and artificial selection. *Proceedings of the Cambridge Philosophical Society*, 1927, *v. 23*, p. 607.

Haldane, J. B. S. A mathematical theory of natural and artificial selection. Part V: Selection and mutation. *Proceedings of the Cambridge Philosophical Society*, 1927, *v. 23*, p. 838.

Haldane, J. B. S. *The Causes of Evolution*. London, 1932.

Husson, R. Sur la mesure de l'accroissement des populations. *XVe Congrès International d'Anthropologie et d'Archéologie Préhistorique, Ve Session de l'Institut International d'Anthropologie*, Paris, 1931, p. 201.

Insolera Filadelfo. Su particolari equazioni di Volterra e loro applicazione finanziaria e demografica. *Rendiconti del R. Istituto Lombardo di Scienze e Lettere*, Milano, 1927, p. 247.

Kreis, H. Stabilität einer sich jährlich erneuernden Gesamtheit. *Bulletin de l'Association des Actuaires Suisses*, 1936, *v. 32*, p. 17.

Lotka, A. J. Relation between birth rates and death rates. *Science*, 1907, *v. 26*, p. 21.

Lotka, A. J. Studies on the mode of growth of material aggregates. *American Journal of Science*, 1907, *v. 24*, pp. 199, 375.

Lotka, A. J. A problem in age distribution. *Philosophical Magazine*, 1911, *v. 21*, p. 435.

Lotka, A. J. A natural population norm. *Journal of the Washington Academy of Sciences*, 1913, *v. 3*, pp. 241, 289.

Lotka, A. J. The relation between birth rate and death rate in a normal population, and the rational basis of an empirical formula for the mean length of life, given by William Farr. *Quarterly Publications of the American Statistical Association*, 1918, *v. 16*, p. 121.

Lotka, A. J. A simple graphic construction for Farr's relation between birth rate, death rate and mean length of life. *Journal of the American Statistical Association*, 1921, *v. 17*, p. 998.

Lotka, A. J. The stability of the normal age distribution. *Proceedings of the National Academy of Sciences*, 1922, *v. 8*, p. 339.

Lotka, A. J. On the true rate of natural increase of a population. *Journal of the American Statistical Association*, 1925, *v. 20*, p. 305.
Lotka, A. J. The progeny of a population element. *American Journal of Hygiene*, 1928, *v. 8*, p. 875.
Lotka, A. J. The actuarial treatment of official birth records. *Eugenics Review*, 1927, *v. 19*, p. 257.
Lotka, A. J. The spread of generations. *Human Biology*, 1929, *v. 1*, p. 305.
Lotka, A. J. Biometric functions in a population growing in accordance with a prescribed law. *Proceedings of the National Academy of Sciences*, 1929, *v. 15*, p. 793.
Lotka, A. J. The structure of a growing population. *Human Biology*, 1931, *v. 3*, p. 459.
Lotka, A. J. Zur Dynamik der Bevölkerungsentwicklung. *Allgemeines Statistisches Archiv*, 1932, *v. 22*, p. 587; 1933, *v. 23*, p. 98.
Lotka, A. J. Industrial replacement. *Skandinavisk Aktuarietidskrift*, 1933, p. 51.
Lotka, A. J. Applications de l'analyse au phénomène démographique. *Journal de la Société de Statistique de Paris*, 1933, *v. 74*, p. 336.
Lotka, A. J. A historical error corrected. *Human Biology*, 1937, *v. 9*, p. 104.
Lotka, A. J. Population analysis: A theorem regarding the stable age distribution. *Journal of the Washington Academy of Sciences*, 1937, *v. 27*, p. 299.
Lotka, A. J. Quelques résultats récents de l'analyse démographique. *Congrès International de la Population*, Paris, 1937. See also *Journal of the American Statistical Association*, 1938, *v. 33*, p. 164.
Lotka, A. J. The application of mathematical analysis to self-renewing aggregates. *Annals of Mathematical Statistics*, 1939, p. 1.
Lotka, A. J. An integral equation in population analysis. *Annals of Mathematical Statistics*, June 1939.
Moser, C. Über Gleichungen für eine sich erneuernde Gesellschaft mit Anwendung auf Sozialversicherungskassen. *Verhandlungen der schweizerischen naturforschenden Gesellschaft*, Schaffhausen, 1921.
Moser, C. Beiträge zur Darstellung von Vorgängen und des Beharrungszustandes bei einer sich erneuernden Gesamtheit. *Bulletin de l'Association des Actuaires Suisses*, 1926, *v. 21*, p. 1.
Meidell, B. Mathematical statistics: On damping effects and approach to equilibrium in certain general phenomena. *Journal of the Washington Academy of Sciences*, 1928, *v. 18*, p. 437.
Norton, H. T. J. Natural selection and Mendelian variation. *Proceedings of the London Mathematical Society*, 1926, *v. 28*, p. 1.
Risser, R. N. Sur une application de l'équation de Volterra au problème de la répartition par âge dans les milieux à effectif constant. *Comptes Rendus de l'Académie des Sciences*, Paris, November 1920, p. 845.
Schoenbaum, E. Anwendung der Volterraschen Integralgleichungen in der mathematischen Statistik. *Skandinavisk Aktuarietidskrift*, 1924, p. 241; 1925, p. 1.
Schulthess, H. Volterrasche Integralgleichung in der Versicherungsmathematik. *Inaugural-Dissertation*, Bern, 1935.
Schulthess, H. Über das Erneuerungsproblem bei Verwendung eines analytischen Sterbegesetzes. *Bulletin de l'Association des Actuaires Suisses*, 1937, *v. 33*, p. 69.
Wyss, H. Lage, Entwicklung und Beharrungszustand der eidgenossischen Versicherungskasse. *Bulletin de l'Association des Actuaires Suisses*, 1929, *v. 24*, p. 39.
Zwinggi, E. Beiträge zu einer Theorie des Bevölkerungswachstums. *Bulletin de l'Association des Actuaires Suisses*, 1929, *v. 24*, p. 95.
Zwinggi, E. Die Witwenversicheerung als Teil der allgemeinen Alters- und Hinterlassenen Versicherung. *Bulletin de l'Association des Actuaires Suisses*, 1931, *v. 26*, p. 79.
Zwinggi, E. Zum Problem der Erneuerung, Blätter für Vers.- Mathematik und verwandte Gebeite. *Beilage zur Zeitschrift für die gesamte Vers.-Wissenschaft*, 1931, *v. 31*, p. 18.
Zwinggi, E. Zur Methodik der Bevölkerungsvorausberechnung. *Zeitschrift für schweizerische Statistik und Volkswirtschaft*, 1933, *v. 69*, p. 255.

2. Population Growth

Goldziher, K. Beiträge zur Theorie der Vermehrungsformeln. *Aktuárské Vedy*, 1930, *v. 2*, p. 1.

Goldziher, K. Methodische Untersuchungen zu den bevölkerungsstatistischen Grundlagen der schweizerischen Alters- und Hinterlassenenversicherung. *Zeitschrift für schweizerische Statistik und Volkswirtschaft*, 1930, *v. 66*, p. 501.

Krummeich, E. Contribution à l'étude du mouvement de la population. *Journal de la Société de Statistique de Paris*, 1927, *v. 68*, pp. 119, 157, 165, 191, 230.

Kuczynski, R. R. *The Balance of Births and Deaths. Vol. 1, Western and Northern Europe*. New York, 1928; *Vol. 2, Eastern and Southern Europe*, Washington, 1931.

Kuczynski, R. R. Rückgang der Fruchtbarkeit. Entwicklung der Fruchtbarkeit. *Bericht über den XIV Internationalen Kongress für Hygiene und Demographie*, *v. 3*, p. 1472.

Kuczynski, R. R. *The Measurement of Population Growth*. 1935.

Kuczynski, R. R. *Population Movements*. Oxford, 1936.

Landry, A. Taux rectifiés de mortalité et de natalité. *Journal de la Société de Statistique de Paris*, 1931.

Rastoin, E. Analyse et prévision démographiques. *Journal de la Société de Statistique de Paris*, 1932, *v. 73*, p. 367.

Sauvy, A. Sur les taux de stabilisation d'une population. *Journal de la Société de Statistique de* Paris, 1934, *v. 75*, p. 51.

Vianelli, S. Evoluzione economica e demografica negli schemi delle curve logistiche. *Revista Italiana di Scienze Economiche*. Bologna: Maggio–Giugno, 1937, *v. 7*.

Vianelli, S. A general dynamic demographic scheme and its application to Italy and the United States. *Econometrica*, 1936, *v. 4*, p. 269.

3. Demographic Projections

Bowley, A. L. *Estimates of the working population of certain countries in 1931 and 1934*. League of Nations, Economical and Financial Section. Geneva, 1926.

Burgdörfer, F. Die Dynamik der künftigen Bevölkerungsentwicklung im deutschen Reich. *Allgemeines Statistisches Archiv*, 1932, *v. 22*, p. 161.

Cannan, E. The probability of a cessation of the growth of population in England and Wales during the next century. *Economic Journal*, 1895, *v. 5*, p. 505.

Cannan, E. The changed outlook in regard to population, 1831–1931. *Economic Journal*, 1931, *v. 41*, p. 519.

Charles, E. The effect of present trends in fertility and mortality upon the future population of England and Wales and upon its age composition. *Royal Economic Society*, 1935, Memorandum 55.

Cramér, H. Über die Vorausberechnung der Bevölkerungsentwicklung in Schweden. *Skandinavisk Aktuarietidskrift*, 1935, p. 35.

Cramér, H. Mortality variations in Sweden. *Skandinavisk Aktuarietidskrift*, 1935, p. 161.

Depoid, P. Les récentes tendances démographiques dans le monde. *Journal de la Société de Statistique de Paris*, 1937, *v. 78*, p. 4.

Dublin, L. I. The outlook for the American birth rate. *Problems of Population*. The International Union for the Scientific Investigation of Population Problems, London, 1932, p. 115.

Gunther, E. Wert oder Unwert der Vorausberechnung der künftigen Bevölkerung. *Allgemeines Statistisches Archiv*, 1936, *v. 25*, p. 404.

Lotka, A. J. The progressive adjustment of age distribution to fecundity. *Journal of the Washington Academy of Sciences*, 1926, *v. 16*, p. 505.

Mises, R. V. Über die Vorausberechnung von Umfang und Altersschichtung der Bevölkerung Deutschlands. Blätter für Vers.-Mathematik und verwandte Gebiete. *Beilage zur Zeitschrift für die gesamte Vers.-Wissenschaft*, *v. 33*, p. 359.

4. Fertility and Family Composition

Burks, B. S. Statistical method for estimating the distribution of sizes of completed fraternities in a population represented by a random sampling of individuals. *Journal of the American Statistical Association*, 1933, *v. 28*, p. 388.
Fegiz, P. L. I Cognomi di San Gimignano. *Metron*, September 1925, *v. 5*, p. 115.
Geissler, A. Beiträge zur Frage des Geschlechtsverhältnisses der Geborenen. *Zeitschrift des K. Sächsischen Statistischen Bureaus*, 1889, *v. 35*, p. 1.
Gini, C. Sur la mesure de la fécondité des mariages. *Revue de l'Institut International de Statistique*, 1934, p. 40.
Hersch, L. Der Rückgang der Geburtenziffer in Deutschland und seine mathematische Formulierung. *Allgemeines Statistisches Archiv*, 1932, *v. 22*, p. 179.
Lorimer, F. Adjustment for the influence of age composition on estimates by Burks' method of the distribution of sizes of completed fraternities. *Journal of the American Statistical Association*, 1935, *v. 30*, p. 688.
Lotka, A. J. The measure of net fertility. *Journal of the Washington Academy of Sciences*, 1925, *v. 15*, p. 469.
Lotka, A. J. The size of American families in the eighteenth century. *Journal of the American Statistical Association*, 1927, *v. 22*, p. 154.
Lotka, A. J. The geographic distribution of intrinsic natural increase in the United States, and an examination of the relation between several measures of net reproductivity. *Journal of the American Statistical Association*, 1936, *v. 31*, p. 273.
Lotka, A. J. The extinction of families. *Journal of the Washington Academy of Sciences*, 1931, *v. 21*, pp. 377, 453.
Muller, J. H. Human fertility in relation to ages of husband and wife at marriage and duration of marriage. *Annals of Eugenics*, 1931, *v. 4*, p. 238.
Savorgnan, F. La statistica delle nascite secondo l'ordine di generazione. *Revue de l'Institut International de Statistique*, 1933, *part 1*, p. 23.
Wicksell, S. D. Sveriges framtida befolkning under olika förutsättningar. *Ekonomisk Tidskrift*, 1926, p. 91.
Wicksell, S. D. Individens fruktsamhet och släktets förökning. *Ekonomisk Tidskrift*, 1930, p. 29.
Wicksell, S. D. Nuptiality, fertility and reproductivity. *Skandinavisk Aktuarietidskrift*, 1931, p. 125.
Wicksell, S. D. Bidrag till den formella befolkningsteorien. *Statsökonomisk Tidsskrift*, Olso, 1934.
Wicksell, S. D. Fruktsamhet och förökning. *Hygienisk Revy*, Lund, 1936.
Wilson, E. B. Size of completed families. *Journal of the American Statistical Association*, 1935, *v. 30*, *Supplement*, p. 577.

5. The Decrease in Births

Burgdörfer, F. *Sterben die weissen Völker*? München, 1934.
Lotka, A. J. Modern trends in the birth rate. *Annals of the American Academy of Political and Social Science*, November 1936, p. 1.
Muller, J. *Der Geburtenrückgang*, Jena, 1924.
Wolfe, A. B. The population problem since the World War. *Journal of Political Economy*, 1928, *v. 36*, pp. 529, 662; 1929, *v. 37*, p. 87.
Wolf, J. *Der Geburtenrückgang*, Jena, 1912.

6. General Works

Dublin, L. I., editor. The American People: Studies in Population. *Annals of the American Academy of Political and Social Science*, 1936.
Dublin, L. I. and A. J. Lotka. *Length of Life*. New York, 1936.

Knibbs, G. H. The mathematical theory of population, of its character and fluctuations, and of the factors which influence them. *Census of the Commonwealth of Australia*, 1917, *Appendix A, v. 1*.
Landry, A. *La Révolution Démographique*. Paris, 1934.
Lorimer, F. and F. Osborn. *Dynamics of Population*. New York, 1934.
Lotka, A. J. *Elements of Physical Biology*. Baltimore, 1925.
Lotka, A. J. Contact points of population study with related branches of science. *Proceedings of the American Philosophical Society*, 1939, p. 601.
Notestein, F. W. and I. B. Tauber, editors. *Population Index*. School of Public Affairs, Princeton University, and Population Association of America.
Paulinus, G. Prolegomena zu einer Bevölkerungsprognose. *Inaugural Dissertation*, Leipzig, 1934.
Thompson, W. S. and P. K. Whelpton. *Population Trends in the United States*. New York, 1933.

See also the publications of the League of Nations and the Report of the International Population Congress, Paris, July–August 1937, in particular the communications of H. Bunle, P. Depoid, R. Husson, A. Landry, A. J. Lotka, S. D. Wicksell.

Author Index

Subject Index

ISBN 0-306-45927-2
90000
9 780306 459276